ESTIMATING PARENTAGE RELATIONSHIPS USING MOLECULAR MARKERS IN AQUACULTURE

ESTIMATING PARENTAGE RELATIONSHIPS USING MOLECULAR MARKERS IN AQUACULTURE

PAULINO MARTÍNEZ
AND JESÚS FERNÁNDEZ

Nova Science Publishers, Inc.

New York

LIBRARY OF CONGRESS CATALOGING-IN-PUBLICATION DATA

Martínez, Paulino.
   Estimating parentage relationships using molecular markers in aquaculture / Paulino Martínez, Jesús Fernández (authors).
      p. cm.
   ISBN 978-1-60692-140-1 (softcover)
   1. Fish culture--Research--Statistical methods. 2. Animal paternity--Testing--Statistical methods. 3. Genetic markers--Statistical methods. 4. Fishes--Molecular genetics--Statistical methods. I. Fernández, Jesús. II. Title.
   SH153.5.M37 2008
   639.8--dc22   2008034445

Published by Nova Science Publishers, Inc. ✢ New York

CONTENTS

PREFACE

The inference of parentage relationships between individuals from their molecular resemblance represents a useful methodology for genetic improvement in Aquaculture. This makes possible the reorganization of broodstock in groups with low relatedness, as well as the identification of families in selection programs to avoid the harmful effects of inbreeding and to maintain the highest diversity as possible across generations. The choice of the appropriate genetic marker and the statistical methodology is essential to get the best solution for the questions considered. Depending on the availability of previous information on the genealogy, paternity or kinship analysis will be applied. The simplest approach to paternity inference, involves parent identification through the exclusion of the remaining candidates. However, this can lead to more than one solution, solvable after application of maximum likelihood procedures. The ability to identify the parents of each offspring depends on the sampling scenario (number of candidate parents and the fraction sampled), on the potential of the markers used (polymorphism, genotyping errors), and on the conformance to theoretical assumptions of the statistical models applied. A large number of paternity programs are freely available. They display usually complementary performances, and there is not the best software for all situations.

When we deal with a single group of individuals belonging to the same generation or that can not be separated into known generations, the aim is just to estimate the degree of genetic relationship between them, usually expressed as the coancestry coefficient. The basic idea is to determine how much of the molecular similarity (Identity By State) is due to the Identity By Descent (the really important parameter). There are two groups of relationship estimators from the molecular information. One of them includes methods directed at estimating coancestry for only a pair of individuals at a time, usually relaying on the knowledge of the allele frequencies of the reference population. Estimators within this group can be further divided into those called Method of Moments Estimators (MME) and those based on Maximum Likelihood (MLE). The other group of methods uses jointly the information of all individuals to determine the more probable population/familiar structure. They perform an explicit reconstruction of the genealogy leading to the observed population (at least for one generation above). Depending on the type of estimator and the assumptions used in their development, each of them presents some advantages and/or limitations which should be taken into account when choosing the one to use. In the present study, the most relevant estimators are presented and, also, a list of free software available for their application.

Chapter 1

INTRODUCTION

The study of relatedness using molecular markers begins in humans by means of blood groups (Weir et al., 2006) and later in Drosophila applying chromosome polymorphisms (Milkman and Zeitler 1974) and, especially, allozymes (Meagher 1986). With the arrival of recombinant DNA technology, a wide range of molecular markers are developed, which combine the specificity of PCR with the analytical capacity of automatic sequencers for fragment analysis. Then, microsatellite loci, a type of markers ideal for studying parentage relationships, are discovered. The development, in parallel, of specific statistical methods for this purpose (Luikart and England 1999), gives rise to a large amount of theoretical and empirical works in this field, currently constituting a very active area of research (Anderson and Garza 2006; Hadfield et al., 2006; Wang 2006; Weir et al., 2006).

A great deal of statistical developments associated to parentage relationship studies emerge in close connection with forensic genetics and molecular ecology (Weir et al., 2006). Forensic genetics has mostly focussed in paternity analysis and individual identification. Statistical developments for paternity inference proceeded initially from this field (Chakraborty et al., 1988; Pena and Chakraborty, 1994; Jin and Chakraborty, 1995). On the other hand, molecular ecology is more interested in estimating population parameters (gene flow, effective population size, breeding system and care of progeny, etc.) than relatedness between individuals itself (Wilson and Ferguson 2002; Jones and Ardren 2003). In livestock production, the availability of pedigree information has determined that statistical tools for covering its requirements have been mainly taken from forensics. Aquaculture shows particular characteristics, such as the capture of genetic diversity from natural populations for genetic breeding programs, the high fecundity of aquatic species, the complexity of broodstock management and finally, the production of thousands of larvae and fry from different families in individual tanks. This shapes a particular scenario, which requires from specific solutions. The revision of publications on parentage relationships in Aquaculture from databases reveals that most of these studies have been focussed on fish, and only very recently there appear studies on crustaceans and molluscs (Taris et al., 2005; Jerry et al., 2006a, 2006b; Lucas et al., 2006; Johnson et al., 2007). Consequently, the problematic presented in this revision, as well as the references cited, will be mainly focussed on fish farming.

The relevance of parentage relationship analysis in aquaculture is mainly connected with genetic improvement and to a minor extent, with food and environmental safety (Hastein et al., 2001; Liu and Cordes 2004). The high fecundity of aquatic organisms has frequently determined the application of simple mass selection programs resulting in genetic diversity losses and inbreeding depression problems (Hulata 2001; Gjedrem 2005). The identification of familiar relationships is therefore a key point to apply familiar breeding programs, which combine inter and intrafamily genetic diversity components avoiding these harmful effects (Falconer and McKay 1996). Also, quite often, parentage relationships among breeders are unknown, because there is no detailed information about their origin or because they proceed from other companies. Usually, commercial batches show a high relatedness among them, because they have been obtained from a few breeders.

Genetic breeding programmes are being implemented in no more than 40 aquatic species (Gjedrem 2005). For their development, it is necessary to check the families founded in species with planned crosses like turbot, salmonids or catfish. The complex management of larvae and fry at farm facilities, which involves the movement of fishes between tanks and their classification by size to accommodate feeding and growth, increases the probability of error, and consequently that actual families do not correspond to those planned. Also, it is common from a certain size, the mixture of families in the same tank to avoid environmental interactions for heritability estimation or for the selection programme itself (Castro et al., 2004). In this context, the appropriate weight of families regarding the selection intensity applied, requires the knowledge of the relationships of the individuals selected. On the other hand, in species with mass spawning, such as black sea bream, Senegal sole or European sea bass, it is essential the identification of families which are reared together in the same tank, both to ascertain the breeding structure as well as to apply a familiar selection programme (De León et al., 1998; Boudry et al., 2002; Fessehaye et al., 2006; Castro et al., 2006). Furthermore, in these species heritability estimation to find out the best selection strategy also depends upon family identification (Vandeputte et al., 2005; Lucas et al., 2006; Wang et al., 2006). Finally, an important issue related with genetic breeding programs is the maintenance of genetic diversity across generations. In addition to maintaining a broodstock of appropriate size, several authors have proposed the application of methods based in relatedness coefficient to minimize the loss of genetic diversity across generations (Ballou and Lacy 1995; Caballero and Toro 2000; Sekino et al., 2004).

Other collateral issues related with Aquaculture are fish traceability and the control of releases from farms, which can contaminate natural populations (Wilson and Ferguson 2002). The identification of parentage relationships through molecular markers can permit to trace back individuals with market problems (e.g. toxins) or released in natural populations, to breeders in the farm which have produced them (Hastein et al., 2001; Hayes et al., 2005). A similar situation can also occur in fattening farms, which receive fry from other companies, and where the presence of some anomaly or pathology can be trace back to the company of origin through paternity analysis (Martínez et al., unpublished data).

Chapter 2

GENERAL CONSIDERATIONS ON GENETIC MARKERS AND THEIR POTENTIAL FOR RELATEDNESS ANALYSIS

2.1. PROPERTIES OF GENETIC MARKERS

The study of parentage relationships through molecular markers is based on the genetic similarity existing between individuals, in turn determined by Mendelian genetic transmission laws (Figure 1). Thus, a parent shares an allele *identical by descent* (IBD) with its offspring, and the allocation of parents to a particular offspring is known as paternity inference. Therefore, for this analysis, it will be necessary individuals of two consecutive generations, the parental and the filial ones. On the other hand, the study of kinship between whatever two individuals (dyad) pertaining or not to a particular generation, will be based on the proportion of IBD alleles shared between them. So, full-sibs share on average one IBD allele, though at some loci they will not share any allele (25%) and in another 25%, both alleles will be IBD. Statistical procedures for allocating parents (paternity analysis) will differ from those devoted to estimate the relationship at dyads (kinship/coancestry analysis), and will be treated in two different sections in this revision. On the other hand, genetic resemblance between individuals will be estimated from the allelic similarity in the loci analyzed, and this will be obtained through genotyping. Consequently, two alleles will be *identical by state* (IBS), if they show the same genotypic value, measured for example in base pairs (bp) at microsatellite loci, or as relative electrophoretic migration at allozymes. Two individuals can share IBS alleles in a particular population, though they are not related, depending on the frequency of this allele in the population under study. The key for paternity, and especially for kinship analysis, will be on the 'goodness' of the statistical methods for estimating IBD from IBS and the allelic frequencies of the population studied (Blouin 2003; Jones and Ardren, 2003; Weir et al., 2006).

In this context, the choice of the best genetic marker for achieving the appropriate power is essential. This will be dependent on their polymorphism, accurate genotyping, technical drawbacks, expression model and cost. As a rule of thumb, the higher the polymorphism of a locus, measured as the number of alleles or as the expected heterozygosity, the greater its power for the analysis. Genotyping accuracy is other essential element, since mistakes can lead to false exclusions in paternity analysis or to decrease genetic similarity at specific dyads when estimating kinship (Sancristobal and Chevalet 1997; Hoffman and Amos 2005; Castro et al., 2006; Johnson et al., 2007). The model of expression is other relevant point, since with dominant markers it is not possible to distinguish between homozygous or heterozygous individuals of dominant phenotype, and, consequently, the capacity of estimating relationships notably decreases. Codominant markers can be up to one order more resolutive than dominant ones (Gerber et al., 2000). Finally, cost and technical drawbacks are usually two issues closely connected, the later being related with the amount of genotyping errors.

	M-1	M-2	M-3	M-4	M-5	M-8	M-14	M-125
MOTHER	165183	095097	165167	090134	147151	192198	116138	132136
FATHER	173181	095095	165165	116120	149151	192194	132132	120136
O1	173183	095095	165167	120134	147151	192198	116132	136136
O2	165173	095097	165165	120134	147151	194198	116132	136136
O3	173183	095097	165167	090120	151151	194198	132138	136136
O4	173183	095097	165165	120134	147151	192192	116132	136136
O5	173183	095095	165165	120134	147151	192198	116132	136136
O6	173183	095097	165165	090116	149151	192198	132138	120136
O7	165173	095095	165165	120134	147151	194198	116132	120132
O8	165173	095097	165165	090120	151151	192194	132138	120132
O9	165173	095097	165165	120134	147151	192192	132138	136136
O10	173183	095095	165167	090116	149151	192194	116132	120132
O11	167169	107125	177179	090122	147151	198202	132132	120128

Notice that all offspring share an allele with their parents at all loci, and that full-sibs can share 0, 1 or 2 alleles per locus.
Genotypes are given in base pairs (165183: heterozygote for 165 and 183 bp alleles)
Microsatellite (M) loci are presented in the upper row from M-1 until M-125
Mendelian incompatibilities with one or both parents of the false offspring O11 appeared marked in yellow and red, respectively.

Figure 1. Microsatellite genotypes in a family from a breeding selection programme in turbot (*Scophthalmus maximus*).

Since their discovering, microsatellites have been the most popular markers used for parentage analysis (Hulata 2001; Wilson and Ferguson 2002; Jones and Ardren 2003; Chistiakov et al., 2006). Their high polymorphism, codominant expression, technical simplicity and reliability, as well as their genomic abundance and ubiquity, have determined their appropriateness for this purpose. However, microsatellites are not free from important drawbacks, including their development, sometimes associated to tandem repeat enriched library construction (Zane et al., 2002), whose efficiency is usually below 5% (Carreras-Carbonell et al., 2004; Pardo et al., 2007). Alternatively, it is possible to cross-amplify microsatellites from close related species, though success is highly variable, and sometimes the effort and time consumed is finally greater than constructing enriched libraries. In fish, a high interspecific microsatellite conservation has been observed in salmonids (Angers and Bernatchez 1997), important in sparids (Brown et al., 2005; Pinera et al., 2006; Castro et al., in press), but very low in pleuronectids (Bouza et al., 2002; Castro et al., 2006) and in gadids (Castro *et al.*, unpublished data).

Other relevant problem for parentage analysis associated to microsatellites, is the presence of genotyping errors (Marshall et al., 1998; Castro et al., 2004, 2006; Hoffman and Amos 2005; Pompanon et al., 2005). Genotyping errors are due to the presence of adjacent alleles differing by 1 bp; to "dropping out", related to misamplification of some alleles, particularly the longest ones in heterozygotes of divergent allelic sizes; to confusion with internal standard peaks in automatic sequencers with only one fluorophore; and to stuttering, related with strand slippage during PCR reaction, more common in dinucleotide than in longer microsatellite motifs (Figure 2). All these sources of error can be increased by the tendency of Taq polymerase to add an extranucleotide after each PCR cycle appearing an echo pike/band at each peak produced in the reaction (Pompanon et al., 2005).

Null alleles are due to the presence of mutations in the regions where primers anneal and can be present at high frequencies in some loci (Figure 3; Dakin and Avise 2004; Jerry et al., 2004; Castro et al., in press). This annealing failing leads to non-amplification of the allele, being then genotyped as homozygote, when actually is a heterozygote genotype. Finally, though there are unstable microsatellites with mutation rates above 10^{-2} (Primmer et al., 1998; Bacon et al., 2001), these usually range between 10^{-3} and 10^{-5} (Schlotterer 2000), which represents a negligible source of mismatches in the context of parentage analysis (Jones and Ardren 2003).

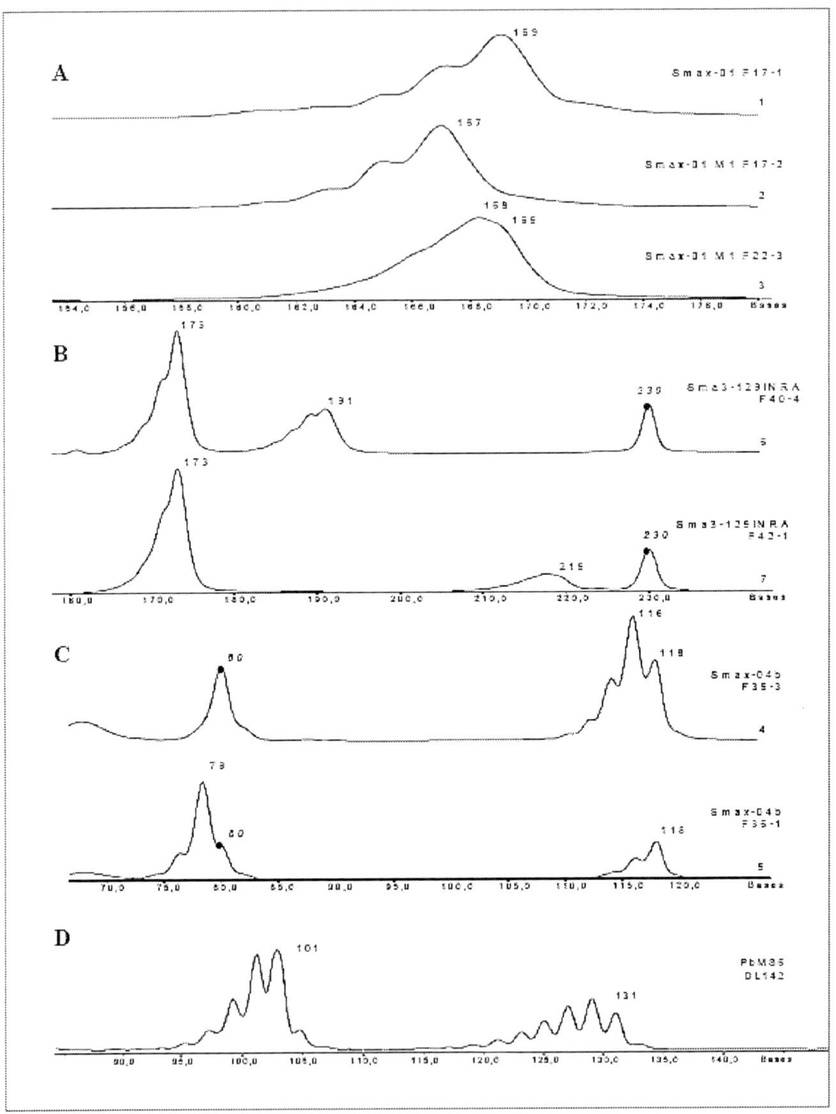

Modified from Castro et al., (2005)

A) Intermediate alleles; B) Drop out; C) Confusion with an internal marker; D) "Stuttering"

The three first cases proceed from turbot (*Scophthalmus maximus*) and the last one from black sea bream (*Sparus aurata*). This one has been obtained by cross-amplification from red sea bream (*Pagellus bogaraveo*).

Figure 2. Sources of genotyping error at microsatellite loci.

FAMILY 1	M-1	M-2	M-3	M-4	M-5	M-6	M-7	M-8	M-9
FATHER	169169	095095	165165	090090	147147	092096	132136	171189	097097
MOTHER	165167	125137	165165	116120	147147	088092	136144	171177	099103
O-1	165169	095125	165165	090116	147147	092092	132136	171189	097103
O-2	167169	095125	165165	090116	147147	092096	136144	171189	097103
O-3	167169	095137	165179	090116	147147	092096	136136	171177	097103
O-4	165169	095137	165165	090120	147147	088096	132136	171177	097103
O-5	165169	095137	165165	090120	147147	088092	136136	171177	097099
O-6	167169	095137	165165	090120	147147	092092	132136	177189	097103

FAMILY 2	M-1	M-2	M-3	M-4	M-5	M-12	M-125	M-129	M-152
Not analyzed	*167/*	*095119*	*179/*	*088090*	*151/*	*092/*	*128132*	*167/187*	*097405*
MOTHER	152169	095109	165179	116122	147147	090092	128148	173173	095099
O1	152167	095119	165179	088116	147151	090106	128132	187187	095097
O2	167169	095095	165179	090122	147151	092098	128148	167173	095105
O3	152167	109119	165179	090122	147151	092098	128128	167173	097099
O4	152167	095119	165179	088116	147151	090106	128132	187187	095097
O5	152167	095095	179179	088122	147151	092098	128148	187187	095105
O6	167169	095109	179179	088122	147151	090098	132148	173187	099105

Microsatellites (M) analyzed in the upper row
Parent-offspring incompatible genotypes highlighted in red
Family 2 illustrates the possibility of parental reconstruction starting from offspring data (italics genotypes).

Figure 3. Mutation (family 1) and null alleles (family 2), as a source of paternity incompatibility in turbot .

Other genetic markers used for parentage relationship inference are the AFLPs ('Amplified Fragment Length Polymorphism'; Gerber et al., 2000; Ezaz et al., 2004), and more recently the SNPs ('Single Nucleotide Polymorphism'). SNPs are due to nucleotide substitutions (Anderson and Garza 2006; Weir et al., 2006), while AFLPs detect a more complex source of molecular variation. SNPs are codominant markers between 10-20 times more frequent across the genome, with higher genotyping accuracy and 5 fold lower cost than microsatellites (Hayes et al., 2005). However, they are usually biallelic, and consequently, exhibit a lower polymorphism than microsatellites. SNPs represent a sound alternative to microsatellites in those species where large genetic resources are available

(Anderson and Garza 2006; Weir et al., 2006). Though genomic resources are being largely increased in the last years in aquaculture species, only in salmonids and in some other model species SNPs have been deposited in public databases. Finally, AFLPs have been occasionally used for parentage analysis in aquatic species. Their main advantage is the low cost per locus and their easy development, especially valuable in species with low genomic resources. However, they are essentially dominant and show lower genotyping confidence than SNPs and microsatellites. Specific statistical methods have been developed for kinship analysis with AFLPs (Gerber et al., 2000), and their performances have been evaluated in comparison with microsatellites in aquatic species (Ezaz et al., 2004). These authors have emphasized that a selection of AFLP loci, where the dominant allele is at frequencies between 0.1 and 0.4, would result in a significant improvement for their allocation capacity.

Most studies on parentage relationships in Aquaculture have been addressed with microsatellite loci. In spite of the potential of SNPs, it will take several years to see their wide application in Aquaculture. Therefore, most examples, references and statistical tools across this revision will be referred to microsatellites.

2.2. STATISTICAL POWER OF GENETIC MARKERS FOR ESTIMATING PARENTAGE RELATIONSHIPS

Though the potential of markers for parentage analysis is closely associated with their polymorphism (number of alleles, heterozygosity), there are more accurate statistical approaches for its assessment. This is a central question to obtain the most powerful set of markers involving the minimum effort for the analysis (Wang 2006). The most popular procedures for this aim are associated with paternity inference derived from forensic genetics (Chakarborty et al., 1988). In this sense, it has been defined the potential of exclusion of a particular locus, as its capacity for excluding a false parent taken at random from the reference population (usually the population of potential parents), when the other parent is unknown (Excl1) or when it is known (Excl2). The Excl2 probability will be always higher than Excl1, since Mendelian restrictions are higher (Table 1). The analysis is based on the calculus of the proportion of genotypes impossible for the true parent. The probability of exclusion of a locus will be equal to the sum of the frequencies of all incompatible genotypes averaged over all possible parents (Chakraborty et al., 1988; Sancristobal and Chevalet 1997). The sampling of genotypes for its calculus is performed using allele frequencies of the reference population assuming Hardy-Weinberg equilibrium. The combined exclusion potential over several loci ($Pe(C)$) is obtained as the product of the probabilities of assignment ($1 - Pe$) for the L loci analyzed assuming independent segregation, according to the following equation:

$$P_E (C) = 1 - \prod_{l=1}^{L} (1 - Pe_l)$$

where Pe is the exclusion probability for the lth locus (Marshall et al., 1998).

However, the probability of exclusion exhibits certain limitations, because it is averaged at sample level and simulation is based in a predefined family representation (Taggart 2007). Additionally, the simulation is performed with a limited number of offspring and does not take into account the existence of genotyping errors and mutations at specific loci (Wang 2006). Finally, the method is based on paternity exclusion and does not consider neither other possible relationships, nor the genotypic information of compatible parents (Wang 2006).

Table 1. Genetic diversity estimators and exclusionary power of 10 microsatellite loci from European sea bass (*Dicentrarchus labrax*)

Locus	Na	He	PIC	Excl1	Excl2
Labrax3	28	0.881	0.870	0.625	0.769
Labrax6	10	0.683	0.635	0.269	0.442
Labrax8	25	0.851	0.838	0.563	0.723
Labrax13	31	0.921	0.914	0.727	0.841
Labrax17	18	0.855	0.837	0.548	0.710
Labrax29	31	0.876	0.866	0.622	0.767
Dla11	16	0.743	0.724	0.381	0.569
Dla12	22	0.920	0.912	0.716	0.835
Dla20	53	0.940	0.935	0.783	0.877
Mean/Total	26	0.852	0.837	0.99978	0.999997

Bouza et al., (unpublished data)
Na: number of alleles
He: expected heterozygosity
PIC: polymorphic information content
Exc1: exclusionary potential 1 (when no parent is known)
Exc2: exclusionary potential 2 (when one parent is known)

Goodnight and Queller (1999) proposed a more powerful method fitted to each relationship context analyzed, though computationally more complex. The method evaluates the potential of a set of loci to discriminate between two alternative hypotheses (primary and null), based in the distribution of the probability ratio between the primary and the null hypotheses obtained through simulation starting from allele frequencies of the reference population. Wang (2006) went forward and introduced the possibility of genotyping errors and mutations, and used a more rapid and accurate method to determine the statistical confidence and potential without a simulation support. Finally, Taggart (2007) has developed an exhaustive and efficient algorithm to evaluate the discriminatory power of a set of markers at family, group or sample level.

Chapter 3

GENERAL CONSIDERATIONS ON STATISTICAL METHODOLOGY

3.1. MAXIMUM LIKELIHOOD ESTIMATION

Parametric statistics assumes that results from a particular process (IBS for markers in this context) are samples from a random variable with a known distribution (the model), which depends on one or several parameters θ (IBD and the allelic frequencies of the reference population). When those parameters are known a density function $f(x \mid \theta)$ can be constructed and thus, we may *a priori* calculate the probability of obtaining a particular value. However, when the estimation is pursued (like determining the degree of relatedness from molecular data), parameters are unknown and we actually take samples of observations to infer their value. Now, we have a fixed data set (X_0: IBSs) and we can define the likelihood of a particular value of the parameter (e.g., full-sibs) as the probability of obtaining the results if the true value of θ were the one considered, $f(X_0 \mid \theta)$. Likelihood function $l(\theta \mid x)$ o $l(\theta)$ is the opposite concept to density function, as it yield the probability of different values of the parameters for a fixed set of observations.

Likelihood allows for the comparison between alternate hypothesis about the true value of the parameters (for example, H_1: one of the individuals being parent of the other, against H_2: they are unrelated). Consequently, the parameter's value with the highest likelihood will correspond with the more probable alternative. Commonly, comparisons are made through the likelihood *ratio*, so the interpretation is simpler, since $l(\theta_1) / l(\theta_2) = 3$ will mean that θ_1 value will be three times more likely than its alternate. Usually, the logarithm of the likelihood is computed [$L(\theta) = \ln l(\theta)$].

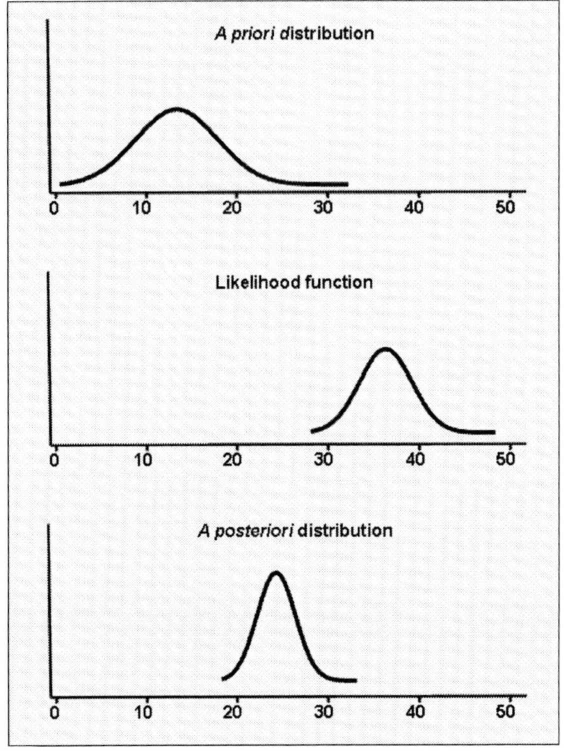

a)

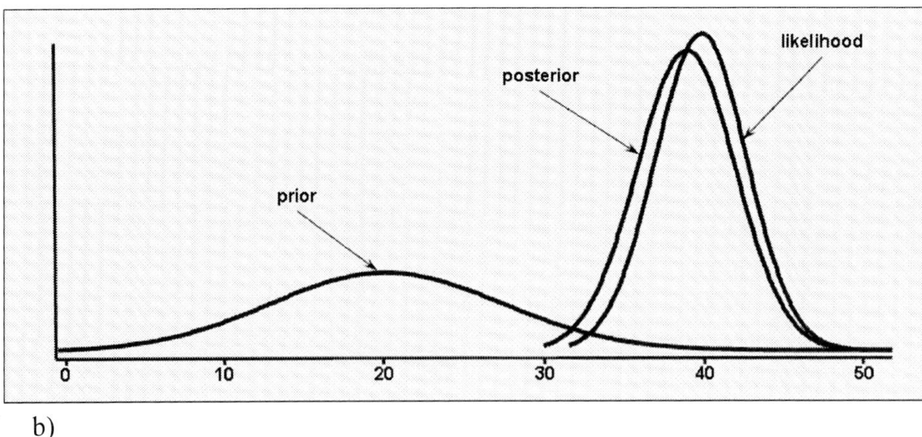

b)

Figure 4. a) Posterior distribution as a combination of the previous information (*a priori*) and the information of data (likelihood). b) With scarce previous information (little informative prior) posterior is very similar to likelihood.

To find the maximum likelihood estimator of the parameter we can calculate the likelihood of a broad range of values (ideally the whole range) and use the one with the highest likelihood. But if the likelihood function can be constructed and derived, we know that the maximum value can be obtained by equating the first derivative to zero and solving for the parameter (if the maximum is not at the extreme of the distribution).

To obtain a measure of the accuracy of our estimate confidence intervals can be constructed around the maximum likelihood estimator. More thoroughly explanation on this methodology and its comparison with other types of estimator can be found in any general book of statistics.

3.2. BAYESIAN INFERENCE

If we consider parameters as random variables which depends on fixed observations, then its distribution $f(\theta \mid X_0)$ can be obtained. The conditional probability of event A given B is the probability that both events occur, provided that event B has actually occurred. Bayes theorem allows for the calculation of such probability following the expression

$$p(A \mid B) = \frac{p(B \mid A) \cdot p(A)}{p(B)}$$

If we translate this formula to the estimation of parameters from the sampled data we have:

$$f(\theta \mid X_0) = \frac{f(X_0 \mid \theta) \cdot f(\theta)}{f(X_0)} = \frac{f(X_0 \mid \theta) \cdot f(\theta)}{\int f(X_0 \mid \theta) \cdot f(\theta) d\theta}$$

A careful reading of this expression tell us that the final or *a posteriori* distribution depends on the probability of the data given the value of the parameter (the likelihood if we consider it as a function of θ) and the *a priori* distribution of the parameter itself. This in turn is related with the previous information about the process we have. The denominator does not depend on θ, and does not affect the parameter estimation. Thus, the posterior distribution is in some way the product of the prior by the likelihood and it will range between them, as shown in Figure 4a.

The main objection against the Bayesian inference is that the result depends on previous assumptions (which may include the researcher prejudices) through the *a priori* probability. To avoid this, a non-informative *a priori* distribution may be used (e.g., a uniform distribution along the definition range of the parameter) and the results will only depend on the likelihood. Notwithstanding, this feature is appealing because it permits the Bayesian methodology to easily process information in a sequential pattern. If after a first estimation we obtain a second set of independent observations, we can repeat the analysis using as *a priori* the posterior distribution obtained with the first sample and the likelihood of the second one.

Another advantage is that estimation consists of a whole distribution, instead of single point estimation and a confidence interval as it happens with maximum likelihood estimation. When a single value is required, commonly the mode of the posterior distribution is the chosen value (it is the most probable value). If we use a criterion minimising the risk of the estimator of giving erroneous values (usually a quadratic function) then the chosen value as a punctual estimator will be the mean.

The smaller the amount of initial information, the closer the posterior distribution will be to the likelihood function, and its mode will be more similar to the maximum likelihood estimate (see Figure 4b). When sample sizes are large, Bayesian estimators converge to maximum likelihood estimators, because the importance of the data is greater than the previous information we may have. Moreover, the mean and the mode of the posterior distribution will be quite close due to the asymptotic normality of this distribution.

Deeper insights on the Bayesian methodology can be found in the book by Bernardo and Smith (1994).

3.3. MCMC Methods

Many times the process we want to characterize depends on more than one parameter and, consequently, the posterior distributions also do (joint posterior distribution). But obtaining an estimate for each parameter requires the calculation of the posterior distribution depending on a single parameter, integrating for the rest (marginal posterior distribution). This integration is difficult to perform (if not impossible), so we have to implement algorithms which allows for the construction of the marginal distributions without an explicit integration (Gelfand and Smith 1990). All MCMC methods (Markov Chain Monte Carlo) arise from the Metropolis-Hastings algorithm, the oldest one, being the most popular the

Gibbs sampling. Markov chains are stochastic successions of events or states in which the probability of a particular state is only function of the previous state. *Gibbs sampling* is based on this kind of chains and basically consists in correlatively sampling from the complete distribution of each parameter, in cycles, converging this chain to the marginal distribution for each of them.

MCMC methods are also useful in searching across a huge space of feasible solutions (for example, all possible combinations of individuals in the population into full-sibs groups) to find the one with highest probability.

3.4. SIMULATION

As we pointed out in the previous section, some operations in the field of the estimation of genetic relationships are hardly manageable in an arithmetic way and require the use of other approaches. In these situations one of the available tools is the computer simulation. Some problems where simulation can be implemented, as explained later, include the determination of significance values for likelihood ratios, the calculation of confidence intervals for some estimators or the establishment of thresholds between adjacent relationship classes (e.g., separation between half-sibs and full-sibs).

The basic idea consists in generating a series of virtual individuals carrying alleles in each locus with a probability equal to the frequency, known or estimated, in the same population. From them, dyads (or groups) of offspring are generated following the Mendelian transmission rules and, therefore, relationship between any pair of individuals are known. From the alleles at the marker loci we can estimate relationships and, with many dyads of the same type, construct a distribution of the estimator which allows for the calculation of confidence intervals and other parameters.

If available, it would be better to simulate virtual parents from the genotypic or haplotypic frequencies, because in this way the possible disequilibrium between or within loci can be taken into account in the process.

Chapter 4

PATERNITY ANALYSIS

4.1. METHODS FOR PARENTAGE ALLOCATION

The fact that each parent necessarily shares one allele with its offspring makes possible an efficient approach to paternity analysis based on exclusion. This is not possible when the relationship between two individuals not directly related through consecutive generations is evaluated. This circumstance also permits to go back from offspring to dead or non-sampled parents for genotyping reconstruction, and makes also possible to estimate the minimum number of parents responsible of a particular progeny. But, as in kinship analysis, efficient Bayesian and maximum likelihood methods have been developed for paternity inference.

4.1.1. Allocation by Exclusion

At first sight, this is the simplest and most appealing solution for parentage allocation (Figure 5). In summary, the true parent is identified by excluding the remaining candidates based on the detection of Mendelian incompatibilities at the loci used. Obviously, the potential of exclusion will be dependent on the number of candidates, the number of loci and their power of exclusion (Wilson and Ferguson 2002; Jones and Ardren 2003). The higher the number of candidate parents, the larger the number of loci necessary for excluding all false parents. However, this relationship is not linear, and the inclusion of an additional locus compensates for a proportionally greater increase of new candidates (Bernatchez and Duchesne 2000). Finally, if one parent is known the allocation of the other parent will be much more straightforward (Marshall et al., 1998).

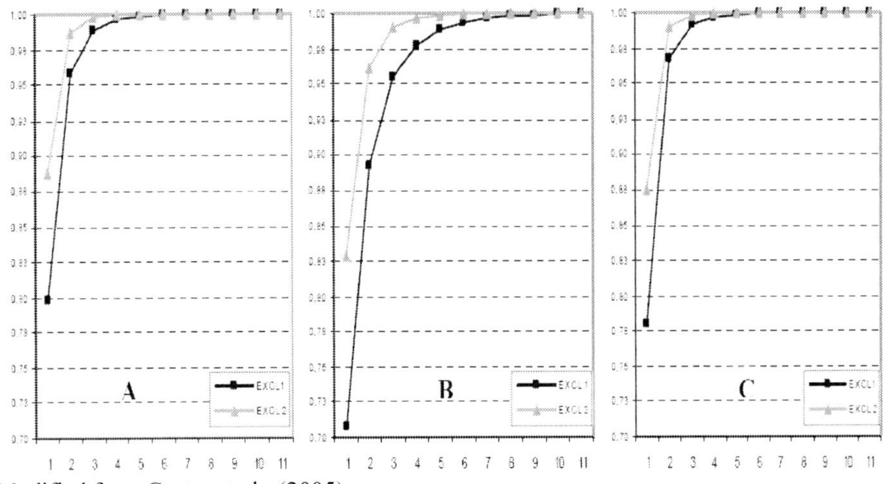

Modified from Castro et al., (2005)

A: black sea bream (*Sparus aurata*); B: turbot (*Scophthalmus maximus*); C: Senegal sole (*Solea senegalensis*)

Abscises: number of loci analyzed; Ordinates: exclusionary probabilities when no parent is known (Exlc1) and when one parent is known (Excl2).

Figure 5. Exclusionary potential variation of microsatellites in three Aquaculture species.

However, this approach presents some drawbacks and can perform better at specific scenarios. Firstly, and depending on the number of candidates and loci managed, it is possible that not all candidates but the true parents are excluded (Wilson and Ferguson 2002). The application of the minimum possible number of loci to alleviate the effort and cost in Aquaculture, increases undefined situations, appearing trios or higher numerous groups of putative parents (Estoup et al., 1998; Castro et al., 2004, 2006, 2007; Johnson et al., 2007). Secondly, the strict exclusion principle is valid when no genotyping errors or null alleles exist at the markers used, which is not a realistic assumption. The percentage of genotyping errors for microsatellite loci, the most popular markers for paternity inference, ranges between 0.5-2% (Estoup et al., 1998; Hoffman and Amos 2001; Castro et al., 2004, 2006, 2007; Pompanon et al., 2005). The proportion of null alleles is variable, but can affect to 10% of the loci, and though their frequencies are usually below 10%, they can reach 20% or even higher values (Dakin and Avise 2004; McCartney et al., 2004; Castro et al., 2007). Both circumstances would lead to false exclusions, but also, though less frequently, to erroneous assignments. The existence of genotyping errors and null alleles leads to an epistemological paradox of the exclusionary approach. The increase in the number of markers, necessary to gain allocation potential, unavoidably leads to a higher number of false exclusions (Marshall et al., 1998; Duchesne et al., 2005). The solution is to introduce certain flexibility in the system permitting a predetermined number of incompatibilities (Vandeputte et al., 2006; Taggart 2007). Obviously, this solution decreases the potential of exclusion. Additionally, and since the tolerance is homogeneously applied at all managed loci, the existence of an important error rate at specific loci, will lead to a higher flexibility in the whole system, increasing incorrect allocations. In this sense, the previous checking of loci for null alleles and genotyping errors using family or population data will facilitate the adjustment of error rate at the sample managed (Marshall et al., 1998; Taggart 2007; Castro et al., 2007).

4.1.2. Maximum Likelihood Allocation

Maximum likelihood methods do not intend to exclude false parents, but allocating the most probable one taking into account the genetic frequencies of a reference population in addition to the genotypes of candidate parents and offspring (Meagher and Thompson, 1986; Marshall et al., 1998; Jones and Ardren, 2003). In this way, the uncertainty associated to several compatible parents is overcome (Anderson and Garza 2006). On the other hand, the existence of genotyping errors will not preclude the identification of the true parent, which can still be the most probable solution. For this analysis, as in all methods of paternity inference, it should be considered if the other parent is known, and if one is searching for the most probable single parent or couple. The likelihood of each candidate parent (or couple) of the offspring tested will be obtained as the log of the likelihood ratio ('LOD score') of that candidate being the true parent respect to one individual taken at random from the reference population. These random individual is obtained from allelic frequencies of the reference population assuming HW equilibrium. This likelihood will be estimated as the proportion of alleles shared per locus. As pointed out by Jones and Ardren (2003) the method will introduce a bias, since homozygotes will have a higher probability of sharing the compatible allele with its offspring. If loci segregate at random, a usual and reasonable assumption in spite of no mapping information available, it is possible to obtain a combined estimation over all loci analyzed by summing the LOD scores at individual loci.

Knowing the most likely parent is not enough to soundly establish paternity. It is necessary to know the statistical confidence, since there can be putative parents with close likelihoods. A relevant contribution to solve this question has been the simulation approach developed by Marshall et al., (1998), applied with some modifications by other authors (Goodnight and Queller 1999). Essentially, the method estimates the difference in LOD scores between the two most probable candidates (Δ) and establishes confidence thresholds through simulation from allele frequencies of the reference population (Table 2). An alternative Bayesian approach which takes into account the information of all compatible pairs has been proposed by different authors (Cercueil et al., 2002; Signorovitch and Nielsen 2002). However, the interpretation of the posterior probabilities is not clear, because no methods to obtain critical values for significance have been suggested.

Maximum likelihood allocation can be categorical or fractional (Jones and Ardren 2003). In the former, the goal is to identify the true parents of the offspring tested. Fractional allocation distributes the probability of each offspring among all parents with LOD score > 0. This approach is meaningful when one is interested in allocating the parentage success of groups of individuals and can perform better than the categorical one for this task (Nielsen et al., 2001; Duchesne et al., 2002; Araki and Blouin 2005). This approach can have some interest in species with mass spawning in Aquaculture when breeders of different ages or origins are managed. However, the categorical allocation is much more widely applied, since the applications in this field are mostly focused at individual level.

Table 2. Paternity assignment output from CERVUS following a maximum likelihood approach

Offspring	P non-exclusion	Candidates	N loci	N incomp	LOD score		Confid
O2	1,59E-04	P1	11	3	2,84E+00	2,69E-01	+
O2	1,59E-04	P2	11	4	2,57E+00	0,00E+00	
O2	1,59E-04	P3	11	2	2,44E+00	0,00E+00	
O3	5,54E-03	P4	11	0	5,22E+00	2,42E+00	*
O3	5,54E-03	P5	11	3	2,80E+00	0,00E+00	
O3	5,54E-03	P6	11	3	2,40E+00	0,00E+00	
O9	5,20E-04	P7	11	1	3,10E+00	2,92E-01	+
O9	5,20E-04	P8	11	3	2,81E+00	0,00E+00	
O9	5,20E-04	P9	11	2	2,09E+00	0,00E+00	

The analysis of three potential offspring (O2, O3 and O9) and 9 candidate parents (P1a P9) based in 11 microsatellite loci is shown

P: probability

N incomp: number of loci incompatible

Confid.: statistical confidence (+ P>80%; * P>95%)

4.1.3. Full Probability Methods

The maximum likelihood models presented do not take into account previous information which could be relevant for parentage assignment. In molecular ecology studies, but also in the aquaculture context, not all individuals have the same *a priori* probability to sire a particular offspring. This information, if appropriately introduced in the system, can improve paternity inference. At farm facilities, there is no random mating and broodstock is structured following spatial-temporal variables. Neef et al., (2001) developed a Bayesian method to introduce *a priori* information, and so, to appropriately weight the search of potential parents to obtain the posterior probabilities. More recently, Hadfield et al., (2006) proposed the so called full probability method in which parent allocation is analyzed simultaneously with the searching of population parameters following an interactive model.

4.1.4. Genotypic Reconstruction of Parents

This approach, rather than identifying putative candidates, intends to reconstruct the genotypes of parents in a highly specific scenario (Jones 2001, 2005). The fact that each parent shares one allele with each offspring makes possible the reconstruction of parental genotypes in full/half-sibs mixtures, when a single female or male is responsible of the progeny. Unfortunately, the method is computationally complex, and no more than six parents of the other sex can be searched. The method works well even when the genotype of the single female (or male) responsible for the progeny is not known (Jones 2005). This procedure could have application in highly specific scenarios in Aquaculture, such as the reconstruction of parents in commercial batches where a single female and several males or *vice versa* are crossed. The estimation of the number of parents responsible of a particular progeny is a relevant point for application of paternity programs, since many of them require both their number and sampled fraction. However, there exist more appropriate methods to solve this question. Thus, the minimum number of parents can be grossly estimated as half the number of alleles of the highest polymorphic locus analyzed. The multilocus minimum method uses the information of all loci and divides all possible haplotypes across loci by 2^L, where L is the number of loci (Fiumera et al., 2001). A computational more sophisticated method has been suggested by Duchesne et al., (2005), which uses the cumulative allocation probability of loci to obtain the proportion of collected and uncollected parents responsible of a particular progeny.

4.2. STATISTICAL METHODOLOGY PROBLEMS

4.2.1. Sampling Scenarios in Paternity Analysis

The strategy for paternity inference is greatly dependent on the sampling scenario, i. e., the number of candidates, the proportion of sampled parents and previous allocation of candidates. The situation in Aquaculture is much simpler than that described in natural populations (Jones and Ardren 2003). The most favourable scenario is when crosses can be planned, such in salmonids, turbot or catfish. In these species, if families are reared in independent tanks, a primary goal is to confirm that the families actually correspond to that planned. The management of a complex broodstock, the movement and classification of fry to encompass growth, the jumping of fish between adjacent tanks, and so on, makes recommendable the checking of short samples per family to confirm their family origin (Castro et al., 2004). If this checking is not carried out, an increasing proportion of errors will be accumulated across the selection programme diminishing the expected progress according to heritability estimates. In these cases, it is not necessary the application of statistical tools, the Mendelian congruence being checked manually, always depending on the amount of data managed. Also, the availability of several individuals per family makes the analysis much more consistent, since genotyping errors or null alleles at specific loci are more easily detected, and the assignation is much more powerful (Jones and Ardren, 2003; Castro et al., 2004, 2006). No more than 2 - 4 microsatellites are usually applied for this task (Castro et al., 2004).

In mass spawning species, the situation is more complex and will be dependent on the number of breeders per tank. Contrarily to what happens in studies in natural populations, here all breeders are usually available for analysis, which makes easier the allocation of parents. Additionally, breeders are subdivided in several tanks and photoperiods, which makes possible to restrict allocation to small subgroups. In these species both parents are unknown, which can suggest allocation by couples, as several statistical programs perform (Cercueil et al., 2002; Kalinowski et al., 2007). Though more powerful, this option should be taken with caution, since genotyping errors at one parent would automatically discard the identification of both parents. A low number of microsatellite markers (3-5) can be enough to achieve a potential assignment above 95% in these cases (Herbinger et al., 1995; Pérez-Enríquez et al., 1999; Hara and Sekino 2003; Dong et al., 2006; Castro et al., 2006, 2007).

As outlined before, acquisition of fish from other commercial sources is a common practice in Aquaculture for renewing genetic diversity and taking advantage of selection gain. Usually, these batches are founded with 1 - 2 females and several males. The identification of kinship among these potential breeders is essential to be incorporated in the selection programme. Here, the disadvantage is the lack of parents, but the low number of founders and the mixture of half/full-sib families make this task affordable through parental reconstruction methods (Jones 2001, 2005).

4.2.2. Treatment of Technical and Genetic Drawbacks of Molecular Markers

At the beginning of this work, it was indicated the types and frequency of genotyping errors associated to each genetic marker. In this section, it will be described and discussed the appropriate way for treating these errors for limiting their impact on paternity inference. Null alleles are one of the main sources of error associated to microsatellites (Wilson and Ferguson 2002) (Figure 6). More than 10% of the microsatellites reported exhibited null allele frequencies around or above 5%, a figure which can seriously compromise paternity inference (Marshall et al., 1998). In some cases, they can be eliminated by redesigning the primers starting from the library clone where they were identified (Dakin and Avise 2004). A first statistical approach to this problem is to use a tool which alerts us about their presence from population data. Null alleles give rise to heterozygote deficit in panmictic populations. Unless this deficit is caused by population structure which affect all loci in a similar way (inbreeding, Wahlund effect), microsatellites with null alleles will exhibit heterozygote deficit (Brookfield 1996). Some programmes designed for paternity inference include algorithms to identify those loci suspicious of harbouring null alleles, to be discarded from the analysis or to be taken with caution for their application (Marshall et al., 1998; Amos 1999). On the other hand, detection of null alleles will be much more accurate if family data are available, or if, as outlined before, the data managed include full-sib families, a situation common in the aquaculture context. In these cases, null alleles will be identified by the presence of homozygote-homozygote mismatches between parent and offspring at specific loci (Figure 3). If enough family data are available, an accurate estimate of null allele frequency can be obtained at each locus (Primmer et al., 1998; Dakin and Avise 2004; Castro et al., 2004, 2006, 2007). An alternative to avoid the exclusion of true parents due to null allele mismatches is the transformation of all homozygote genotypes to heterozygotes for a null allele, at least at those suspicious loci, and especially if an approximation of exclusion for allocation is followed (Danzmann 1997).

It has already been mentioned, that, in spite of exhibiting mutation rates highly above than single nucleotide substitutions false exclusions due to mutations will be exceptional when using microsatellite loci. Thousands of genotypes will not include more than a few incompatibilities due to mutation (Ellegren 2000). Only highly unstable microsatellites could introduce a proportion of mismatches to be taken into consideration (Primmer et al., 1998). However, genotyping errors, as null alleles, can produce much higher impact. Particularly, 'drop out' and 'stuttering' are the cause for systematic errors in microsatellite genotyping, which can notably affect paternity inference (Castro et al., 2004, 2006; Hoffman and Amos 2004; Johnson et al., 2007). Their particular profiles, and especially their iteration, makes easier their detection, especially when family data are available. This obligates to reanalyze electropherograms and facilitates their elimination for further analysis. Most genotyping errors of microsatellites leave a particular signature, which facilitates their detection starting from population data. In this sense, the analysis of expected and observed heterozygote distributions can be used to detect drop out (low frequency of heterozygotes for highly divergent alleles) or stuttering (excess of heterozygotes for shortly distant alleles). The program MICRO-CHECKER (Van Oosterhout et al., 2004) uses this approach to identify and quantify the presence of null alleles, drop out and stuttering.

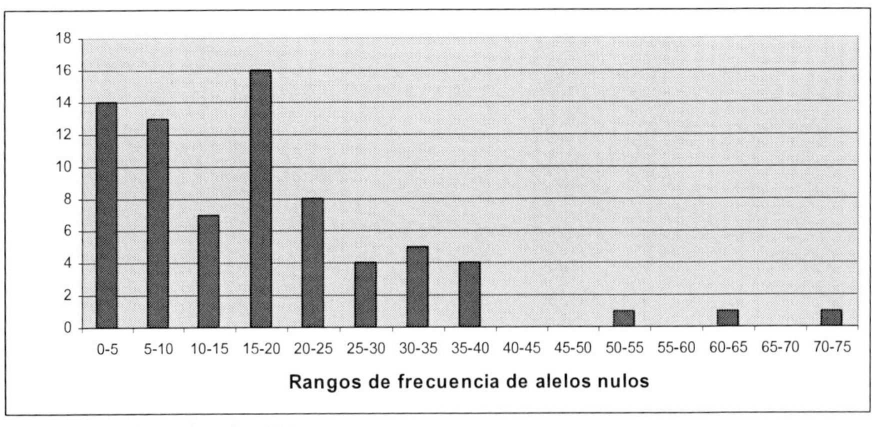

Data from Dakin and Avise (20

Number of loci with null alleles (ordinates) within a particular frequency range (abscises; in percentage).

Figure 6. Null allele frequency at microsatellite loci.

4.2.3. Departure from Theoretical Assumptions

Most approaches for paternity inference, excluding strict exclusion methods, rely on several theoretical assumptions, usually the conformance to HW equilibrium and independent segregation of loci (Wilson and Ferguson 2002). HW is usually assumed when simulating genotypes of parents and offspring for calculating exclusion probabilities or when looking for confidence in paternity inference. The lack of HW equilibrium in the reference population will determine an overestimation of exclusion probabilities, especially when heterozygote deficit is the cause (Estoup et al., 1998; Castro et al., 2006). Also, independent segregation of loci is assumed to calculate the combined potential of exclusion or the combined assignment for paternity. This information is usually unavailable, but when few loci are managed the probability of strong linkage between them which could originate a severe distortion in the analysis is really low. However, linkage has been observed in different studies (Ardren et al., 1999), mostly due to non-independent isolation of clones from enriched libraries. The use of different tandem repeated sequences from the same clone to increase the efficiency of libraries for microsatellite isolation has been observed in several aquaculture species (Iyengar et al., 2000; Bouza et al., 2002; Castro et al., 2006, 2007). The diminishing power for assignation of these linked loci will be associated to the lack of gametic equilibrium, which, in general, will be more intense the closer the linkage between loci. However, if gametic equilibrium exists in spite of linkage, there will not be a departure of parentage assignment from expectations, at least for the whole sample. If linkage phase is known both linked loci would even constitute a superlocus, providing a higher potential for assignment (Jones et al., 1998). However, this is not frequent and this possibility represents more a theoretical exercise rather than a practical tool for paternity analysis. Additionally, though expectations for the whole sample do not change if gametic equilibrium exists, this situation would be variable depending on the particular genotypes of the families analyzed (Anderson and Garza 2006; Castro et al., 2006). Therefore, as a rule it should be recommended the application of unlinked loci for parentage inference.

A final element which notably influences paternity allocation is the presence of a strong family structure among candidates, or especially, if among candidates are included half or full-sibs of the offspring to be allocated (Estoup et al., 1998; Marshall et al., 1998; Castro et al., 2007). This circumstance would be unlike in aquaculture since ages of breeders are usually known, at least when the genetic breeding programme is working. However, it could occur when founding the broodstock from native individuals or when individuals come from unknown commercial source. In these cases, the actual parent could show LOD scores very close to their relatives or even lower when it competes with a full-sib of the offspring analyzed. The best solution would be to increase the number of loci. As a general rule maximum likelihood approaches would be more affected than exclusion ones by this problem (Marshall et al., 1998; Jones and Ardren 2003). These situations would be avoided if breeders were stocked according to minimum relatedness (Castro et al., 2007).

4.3. COMPUTER PROGRAMS FOR PATERNITY INFERENCE AND THEIR PERFORMANCES

A relevant question in paternity inference is the choice of the appropriate program among the many existing ones (Table 3). However, there is not the best software for all situations, and many of them present complementary performances as emphasized across the previous sections in this revision and also by different authors (Kasumovic et al., 2003). The most common scenario in aquaculture breeding programmes is the knowledge of most, if not all, parents of the progeny to be sired. This limits the utility of programmes such as PAPA and PASOS, which offer performances related with the identification of the fraction of parents sampled, more useful for natural populations. Also the requirement of these variables by some programmes, such as CERVUS and FAMOZ (Marshall et al., 1998; Gerber et al., 2000), would not represent a handicap. In fact, as seen later, CERVUS is the most popular program among aquaculture researchers. Also, the problem related with the lack of some parents when an exclusive searching by couples approach is performed (PROBMAX, Danzmann 1997; FAP, Taggart 2007; PAPA, Duchesne et al., 2002) would not represent a drawback in the aquaculture context. In Aquaculture, parental reconstruction provided by GERUD (Jones 2001, 2005) could aid to identify the small fraction of parents which have not been sampled or died (Castro et al., 2004).

Regarding the allocation method, again the availability of all or most parents in aquaculture is a relevant point to take into consideration. The exclusion approaches with variable tolerance implemented in several programs (NEWPAT, Worthington Wilmer et al., 1999; PARENTE, Cercueil et al., 2002; FAP, Taggart 2007) will show here their best performance. Most offspring could be allocated depending on the genotyping accuracy and the number of loci applied. An error rate below 3 - 4% combined with an appropriate tolerance could be enough to solve correctly most paternities (Vandeputte et al., 2006). However, the use of a low number or low polymorphic markers, or the existence of relatedness among candidates could lead to multiple allocation (Castro et al., 2004, 2006). Maximum likelihood methods could perform better when error rate is high and additionally, they could solve multiple allocation paternities. Castro et al., (2007) demonstrated a better performance of the exclusion over maximum likelihood methods when all parents were available in *Sparus aurata*, but after an exhaustive correction of errors and in a scenario with relevant relatedness among parents. This is the worst situation for maximum likelihood methods. More investigation using actual data will be necessary in the future to clarify this point.

Finally, as outlined before, fractional allocation programmes, such as PATRI (Signorovitch and Nielsen 2002) could have some interest to evaluate the relative contribution of different parental groups or strains in species with mass-spawning, since they perform better than categorical allocation approaches for this purpose (Jones and Ardren 2003).

A relevant point to be considered in paternity inference is the treatment of genotyping errors or null alleles, which could induce to false assignations (Jones and Ardren 2003). Regarding this point, the availability of population data would be of great value, since, as outlined before, it would permit the estimation of null allele frequency and the presence of genotyping errors at the loci used. CERVUS or NEWPAT include algorithms to estimate null allele frequencies based on heterozygote deficit, and therefore, users can be alerted against those loci with moderate-high null allele frequencies with reasonable accuracy (Castro et al., 2007). PROBMAX uses a highly conservative option to avoid false exclusions due to null alleles, which implies the recoding of all homozygotes as heterozygotes for a null allele.

The availability of family data is of great value to identify systematic genotyping errors or accurately estimate null allele frequencies (Castro et al., 2004, 2006, 2007). The existence of large progenies in Aquaculture with a limited number of parents permits to identify systematic errors using the progenies analyzed itself, such as implemented in the program FAP (Taggart 2007). This is probably the most appropriate option in an Aquaculture scenario. The remaining programmes use automatic options which treat all mismatches in the same way, irrespective if they are null alleles or genotyping errors. Additionally, they treat all loci in a homogeneous way, leading to an error rate overestimation (Jones and Ardren 2003; Kalinowski et al., 2007). Thus, CERVUS and PARENTE interpret mismatches as replacements by genotypes or alleles taken at random from allele frequencies of the reference population.

Table 3. Paternity inference programs of interest in Aquaculture

Program	Method	Markers	SP	PP	PR	SC	EP	Required information	Management of technical/ genetic sources of incongruence	Comments
PROBMAX[1]	Exclusion	Codominant/ dominant		X				Parent and offspring genotypes. Candidate subgroups if wished	Permits retesting of data assuming SMM errors and recodifing homozygotes for NA	Evaluates exclusion capacity of markers
NEWPAT[2]	Exclusion	Codominant/sex linked	X			X		Offspring and sexed parents genotypes	NA frequency estimates. Exclusion tolerance to be decided by the user	Estimates the minimum number of loci for exclusion
PARENTE[3]	Exclusion	Codominant	X	X		X		Parent and offspring genotypes.	It admits error tolerance to be introduced by the user	Age of candidates can be introduced
FAP[4]	Exclusion	Codominant		X				Parent and offspring genotypes.	Introduces tolerance according to user. Identifies systematic genotyping errors or NAs	It permits to identify the best loci set of loci for assignment
CERVUS[5]	Categorical allocation/ Max likelih	Codominant	X	X		X	X	Offspring and sexed parents genotypes	Proportion of errors defined by the user. It estimates genotyping errors when true parents available	It permits exclusion approach with variable tolerance
PAPA[6]	Categorical allocation/ Max likelih	Codominant		X		X		Offspring and parents genotypes	Error rate introduced by the user using a SMM model.	Identifies the best set of loci for allocation
FAMOZ[7]	Categorical allocation/ Max likelih.	Codominant, dominant and uniparental	X	X		X	X	Parent and offspring genotypes	It can be introduced proportion of errors and deviations from panmixia by the user	Estimates probability of genotypic identity
PASOS[8]	Categorical allocation/ Max likelih.	Codominant	X	X				Parent and offspring genotypes	Transmission errors according to SMM. It admits error tolerance within an exclusion approach	Estimates number of candidates from offspring genotypes

Table 3. Paternity inference programs of interest in Aquaculture (Continued

Program	Method	Markers	SP	PP	PR	SC	EP	Required information	Management of technical/ genetic sources of incongruence	Comments
PATRI[9]	Fractional allocation/ Bayesian	Codominant	X			X		Offspring-parent pairs and sexed parents genotypes	Evaluation of genotyping errors using offspring-parent pairs data	Candidates subgroups can be defined for fractional allocation
GERUD[10, 11]	Parental reconstruction	Codominant			X	X	X	Offspring and known parent genotypes		Estimates minimum parent number

SP: Single parental allocation

PP: Couple allocation

PP: Parental reconstruction of non-genotyped candidates

SC: Statistical confidence for assignment

EP: It provides exclusion probabilities as an estimation of assignment potential per locus

SMM: Stepwise mutation model

NA: Null allele

[1]Danzmann (1997); in: http://www.uoguelph.ca/~rdanzman/software/probmax/

[2]Worthington Wilmer et al., (1999); in:http://www.zoo.cam.ac.uk/zoostaff/amos/newpat.html

[3]Cercueil et al., (2002); in: http://www2.ujf-grenoble.fr/leca/membres/manel.html

[4]Taggart (2006); in: http://www.aqua.stir.ac.uk/rep-gen/Downloads.html

[5]Marshall et al., (1998); in: http://helios.bto.ed.ac.uk/evolgen/cervus/cervus.html

[6]Duchesne et al., (2002); in: http://www.bio.ulaval.ca/contenu-fra/professeurs/Prof-l-bernatchez.html

[7]Gerber et al., (2000); in: http://www.pierroton inra.fr/genetics/labo/Software/Famoz/

[8] Duchesne et al., (2005) ; in : http://www2.bio ulaval.ca/louisbernatchez/downloads.htm

[9]Signorovitch and Nielsen (2002); in: http://www.biom.cornell.edu/Homepages/Rasmus_Nielsen/files.html

[10]Jones (2001); in: http://www.biology.gatech.edu/professors/labsites/jones/parentage.html

[11]Jones (2005); in: http://www.bio.tamu.edu/USERS/ajones/parentage.htm.

In these cases, assignations will be more probable if the error involves a frequent genotype or allele in the reference population. Other programmes, such as PAPA and PASOS calculate a probability depending on a particular mutational model selected by the user, commonly the stepwise model applied to microsatellites. Most programmes based in allocation by exclusion use variable tolerance to be decided by the user (FAP, NEWPAT) and/or permit the recoding of mismatches according to the stepwise mutation model (PROBMAX).

4.4. PATERNITY INFERENCE IN AQUACULTURE

A representative sample of some of the most relevant studies on parentage inference in Aquaculture is presented in Table 4. Firstly, it should be noticed their progressive increase during last years, so in 2006-2007, there appeared as many publications as in previous years jointly. Secondly, though most works have been applied on fish, a remarkable increment on molluscs and crustaceans is evident in the last years. Microsatellites are the markers usually applied for paternity inference in Aquaculture, but in some species with large genomic resources, such as salmonids, SNPs are beginning to be used. The increase in genomic resources (markers, genetic maps) in some aquaculture species is determining the development of more sophisticated molecular tools for throughput parentage allocation by checking exhaustively null alleles, genotyping errors and linkage among markers (Johnson et al., 2007; Li et al., 2007; Castro et al., 2007). CERVUS is the most popular program among aquaculture users, though manual assignation appears still quite common. The new version of CERVUS (Kalinowski et al., 2007) provides several new options over the previous one, such as exclusion approach with tolerance or searching by couples, which could enhance its application. In the future, it will be necessary to evaluate the use of exclusionary-based programmes, such as FAP, specifically designed for aquaculture context. Finally, many publications detected in this survey are related with microsatellite implementation, which ensures a future application of paternity inference and their applications in these species. Heritability estimation, reproductive structure and support in familiar breeding programmes are the most sophisticated applications used in Aquaculture.

Table 4. Representative references for paternity inference in Aquaculture

Species	Taxonomic group	Genetic marker	Program	Application	Reference
Sparus aurata	Fish	Microsatellites	CERVUS	Parentage assignment	Castro et al., in press
Morone spp. hybrids	Fish	Microsatellites	Own program	Heritability estimation	Wang et al., 2007
Oncorhynchus mykiss	Fish	Microsatellites	-	Parentage assignment	Johnson et al., 2007
Penaeus monodon	Crustaceans	Microsatellites	-	Genetic diversity and fingerprinting	Li et al., 2007
Acipenser ruthenus	Fish	Microsatellites	Manual	Gynogenetic evaluation	Fopp-Bayat et al., 2007
Gadus morhua	Fish	Microsatellites	Manual	Parentage assignment	Wesmajervi et al., 2006
Pagellus bogaraveo	Fish	Microsatellites	CERVUS	Parentage assignment	Lemos et al., 2006
Oreochromis niloticus	Fish	Microsatellites	PAPA	Reproductive structure	Fessehaye et al., 2006
Morone sazatilis y Morone chrysops	Fish	Microsatellites	Manual	Heritability estimation	Wang et al., 2006
Haliotis asinina	Molluscs	Microsatellites	Manual	Heritability estimation	Lucas et al., 2006
Fenneropenaeus chinensis	Crustaceans	Microsatellites	CERVUS	Parentage assignment	Dong et al., 2006
Penaeus monodon	Crustaceans	Microsatellites	CERVUS	Parentage assignment	Jerry et al., 2006a
Penaeus japonicus	Crustaceans	Microsatellites	CERVUS	Interfamiliar growth evaluation	Jerry et al., 2006b
Solea senegalensis	Fish	Microsatellites	CERVUS	Parentage assignment	Castro et al., 2006
Salmo salar	Fish	Microsatellites/ SNPs	Manual	Parentage assignment and genetic diversity	Rengmark et al., 2006
Crassostrea gigas	Molluscs	Microsatellites	PAPA	Breeding programme	Taris et al., 2005
Salmo salar	Fish	Microsatellites/ SNPs	Own program	Parentage assignment and traceability	Hayes et al., 2005

Table 4. (Continued)

Species	Taxonomic group	Genetic marker	Program	Application	Reference
Cyprinus carpio	Fish	Microsatellites	Own program	Heritability estimation	Vandeputte et al., 2004
Scophthalmus maximus	Fish	Microsatellites	CERVUS	Parentage assignment	Castro et al., 2004
Oreochromis niloticus	Fish	Microsatellites/ AFLPs	-	Gynogenetic evaluation	Ezaz et al., 2004
Scophthalmus maximus	Fish	Microsatellites	CERVUS PROBMAX	Parentage assignment	Borrell et al., 2004
Scophthalmus maximus	Fish	Microsatellites	CERVUS	Gynogenetic evaluation	Castro et al., 2003
Paralichthys olivaceus	Fish	Microsatellites	Manual	Parentage assignment and genetic diversity	Sekino et al., 2003
Crassostrea gigas	Molluscs	Microsatellites	Manual	Reproductive structure	Boudry et al., 2002
Salmo salar	Fish	Microsatellites	KINSHIP	Parentage assignment	Norris et al., 2000
Clarias gariepinus	Fish	Microsatellites	FAMILY	Breeding programme	Volckaert and Hellemans 1999
Dicentrarchus labrax	Fish	Microsatellites	Manual	Breeding programme	De León et al., 1998
Oncorhynchus mykiss	Fish	Microsatellites	Manual	Breeding programme	Herbinger et al., 1995

Note: The references are a representative sample taken from some relevant publications in Aquaculture.

Chapter 5

KINSHIP ESTIMATION

In the context of genetics, when we talk about kinship or coancestry we refer, in general, to the common origin of alleles carried by two individuals in a particular locus. What we really try to detect or predict is the probability of those alleles being *Identical By Descent* (IBD). This means that both alleles are copies of a single allele in a common ancestor, in opposition to the concept of *Identical By State* or *Alike In State* (IBS or AIS) when alleles are equal although they derive from different alleles in the reference population. A graphical representation of both concepts can be found in Figure 7.

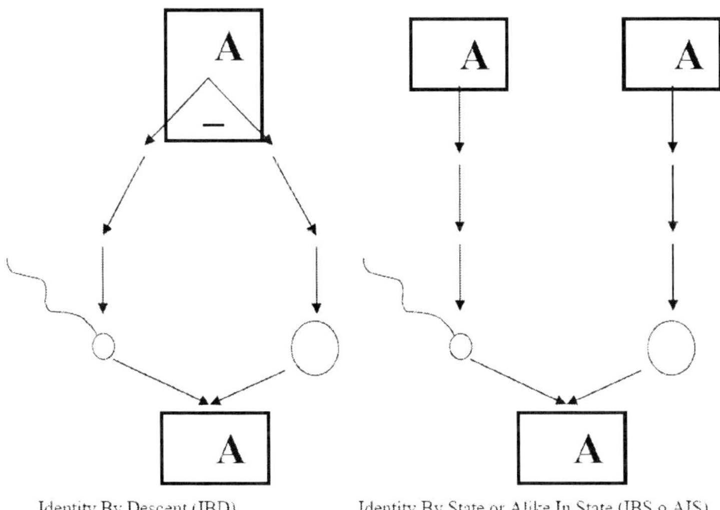

Figure 7. Graphical representation of the difference between alleles identical by descent or by state.

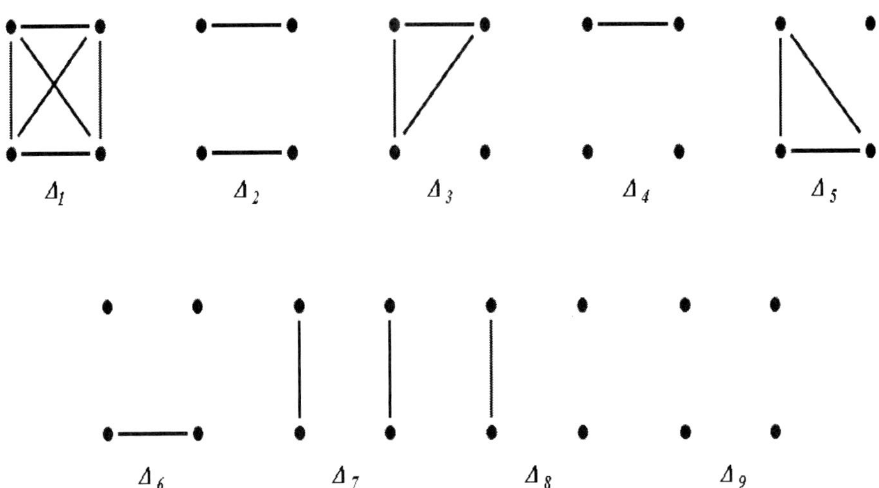

Figure 8. Identity by descent modes between two individuals.

The more general and exhaustive way of describing the relationship between two individuals is in terms of the nine models of identity by descent presented by Jacquard (1974) (see Figure 8). Following this methodology several coefficients Δ = (Δ_1, Δ_2, ..., Δ_9) can be calculated corresponding to the probability of the four alleles of the dyad conforming to that pattern of IBD. Another way of mathematically representing the degree of genealogical relationship is through the probability of the genotypes of two individuals sharing 2 (k_2), 1 (k_1) or 0 (k_0) alleles (Thompson 1991).

Commonly, the degree of genetic relationship between two individuals is summarised using the coancestry coefficient (f or θ, depending on the authors). This was defined by Malecot (1948) as the probability of sampling two alleles identical by descent taken at random one from each individual in the same locus. The coancestry coefficient for non inbred individuals is half the additive genetic correlation or relatedness coefficient (r). Some methodologies have been developed for the estimation of r, but since in most situations the transformation between parameters is straightforward (r = $2f$), we shall refer to f estimators thereafter. When the complete genealogy is known, the coancestry coefficient can be easily calculated following classical rules (Falconer and Mackay 1996) or implementing tabular methods (Emik and Terrill 1949). We should also remember that when comparing two alleles from the same individual the probability of being IBD is the inbreeding coefficient (F), equal to the genealogical coancestry of the parents of that individual.

5.1. MOLECULAR INFORMATION

When the genealogy of the population under study is not available, information from molecular markers developed should be used to estimate kinship. The basic idea is to define a measure of the molecular resemblance between individuals, assuming that those sharing the highest number of alleles for the markers will be those with the highest probability of carrying IBD alleles (Blouin 2003; Weir et al., 2006). Different estimation strategies and procedures exist which can be classified in the groups described below.

5.2. PAIRWISE ESTIMATORS

The main characteristic of these methods is that relationships are determined for just one couple of individuals (dyad) at a time based on their molecular information. Data corresponding to the remaining individuals in the population are not used but for the estimation of allelic frequencies.

5.2.1. Molecular Coancestry

This coefficient arise from the application of Malecot (1948) definition to marker loci. Accordingly, molecular coancestry (f_M) between two individuals is the probability of sampling the same allele (whatever IBD or not) when taking one allele from the same locus of each individual. Molecular coancestry ranges from 0 (when none of the four possible comparisons is between IBS alleles) to 1 (when all four alleles are IBS), with only two intermediate values corresponding to 0.25 and 0.5. The relative contribution of each allele from a particular locus to the molecular coancestry can be calculated separately and then summed up, as shown in Table 5 for a locus with three alleles. If information from several loci is available, molecular coancestry is calculated by averaging their particular value over loci.

The main problem with this measure is that only reflects IBS and, thus, the parameter of interest (IBD) will be inflated.

Table 5. Contribution of each allele to the molecular coancestry between two individuals depending on their genotype

Genotypes	f_{Mi}	f_{Mj}	f_{Mk}	f_M
$A_i A_i , A_i A_i$	1.00	0.00	0.00	1.00
$A_i A_i , A_i A_j$	0.50	0.00	0.00	0.50
$A_i A_i , A_j A_k$	0.00	0.00	0.00	0.00
$A_i A_j , A_i A_j$	0.25	0.25	0.00	0.50
$A_i A_j , A_i A_k$	0.25	0.00	0.00	0.25

5.2.2. Similarity Indices

Originally developed for DNA fingerprinting, these methods are applicable for any electrophoretic pattern where different bands can be observed. If comparing an individual X where n_X bands are detected with another individual Y with n_Y bands, n_{XY} being common, a simple similarity measure can be constructed as

$$S = \frac{2n_{xy}}{n_x + n_y}$$

which corresponds to the Dice's coefficient (Sneath and Sokal 1973). Its value will range from 0 (no bands shared) to 1 (all common bands).

When similarity is calculated based on codominant markers, different modifications have been proposed, for example the one by Li et al., (1993):

$$S = \frac{n_{xy}}{2} \left(\frac{1}{n_x} + \frac{1}{n_y} \right)$$

These authors pointed out that the expectation of such similarity index is

$$E(S) = 2f + (1 - 2f)S_0, \text{ being } S_0 = \sum_{i=1}^{n} p_i^2 (2 - p_i)$$

for a locus with n alleles, where p_i is the frequency of allele i in the reference population. This equation allows for the construction of a first estimator of the genealogical coancestry, expressed as

$$\hat{f} = \frac{S - S_0}{2(1 - S_0)}$$

A different approach was proposed by Bowcok et al., (1994), who suggested an index for multilocus analysis consisting in the total number of shared alleles (a maximum of two per locus) divided by twice the number of genotyped loci.

5.2.3. Methods of Moments Estimators (MME)

In an attempt to estimate IBD from the molecular coancestry, we can use the expectation of that coancestry for a particular allele, that is

$$E(f_{Mi}) = f \cdot p_i + (1 - p_i)p_i^2$$

where p_i is the frequency of the allele i in the reference population and f is the genealogical coancestry (i.e., the IBD probability). The first term in the right hand side of the equation is the probability of finding two alleles i identical by descent, while the second one is the probability of sampling two equal alleles in a population with an allele frequency p_i. Solving for f, a simple estimator of the IBD due to a single allele can be obtained as:

$$\hat{f} = \frac{f_{Mi} - p_i^2}{p_i(1 - p_i)} \quad \text{(Ritland 1996)}$$

Estimates for each allele from the same locus can be averaged in different ways depending on the weighting factor (w_i) assumed (Li and Horvitz 1953; Ritland 1996).

Using a different approach, Lynch and Ritland (1999) developed a coancestry estimator based on the probability of finding a particular genotype in a 'proband' individual (Y with alleles k and m), conditional on the genotype of another 'reference' individual (X with alleles i and j). This value is proportional to the probability of sharing one or two alleles and, therefore, an estimation of the coancestry coefficient can be calculated defining a variable S_{ij}. This will take a value 1 when i and j are equal (irrespective if they are in the same or different individuals) and 0 if they are different. Due to the underlying idea of predicting the genotype of Y through the genotype of X, these methods are commonly called *regression estimators*. In concordance with the methodology used in their construction, these estimators are 'asymmetric' being f_{XY} (X is the reference and Y the proband) not necessarily equal to f_{YX}. The arithmetic mean of both values is used as the estimate of the coancestry.

Related with the previous method, although originally developed for the estimation of average coancestry of groups of individuals, the Queller and Goodnight (1989) estimator is based on the same parameters as the one by Lynch and Ritland (1999). A drawback of this method is that the estimator is not defined for biallelic loci when individuals under study are heterozygous, limiting its use to multiallelic loci.

Table 6a. Different estimators for single locus with their corresponding weights

Assumption	Estimator	Reference
$w_i = p_i(1-p_i)$	$$\hat{f} = \frac{f_{Mi} - \sum_{i=1}^{n} p_i^2}{1 - \sum_{i=1}^{n} p_i}$$	Li and Horvitz (1953)
$w_i = \dfrac{1-p_i}{n-1}$	$$\hat{f} = \frac{1}{n-1} \sum_{i=1}^{n} \frac{f_{Mi} - p_i^2}{p_i}$$	Ritland (1996)
	$$\hat{f}_{XY} = \frac{p_i(S_{jk} + S_{jm}) + p_j(S_{ik} + S_{im}) - 4p_i p_j}{2(1 + S_{ij}) + S_{km}(p_i + p_j) - 8p_i p_j}$$	Lynch and Ritland (1999)
	$$\hat{f}_{XY} = \frac{(S_{ik} + S_{im} + S_{jk} + S_{jm}) - p_i - p_j - p_k - p_m}{2 + S_{ij} + S_{km} - p_i - p_j - p_k - p_m}$$	Queller and Goodnight (1989)
	$$\hat{f}_{XY} = \frac{4P_1 + 3P_2 - 2\left(1 + \sum_{i=1}^{n} p_i^2\right)}{4\left(1 - \sum_{i=1}^{n} p_i^2\right)}$$	Wang (2002)

Wang (2002) developed a regression estimator recalculating the probabilities of finding dyads of individuals with different similarity indices (S = 1, ¾, ½ or 0 following the definition by Li et al., 1993 in the previous section). In his particular formulation, P_1 is equivalent to S = 1 and P_2 to S = ¾. Although in certain situations this estimator coincides with the Queller and Goodnight (1989) and Lynch and Ritland (1999) estimators, it exhibits better performance for some situations, as tested by simulation using actual turbot data (Pino-Querido et al., submitted).

The mathematical expressions of the estimators described above when dealing with single locus information can be found in Table 6a.

Different weights to average the estimates obtained from different loci have been proposed. These will essentially depend on the criteria about the relationship chosen among loci and the weight given to each estimate. Oliehoek et al., (2006), besides including a different weighting factor for each locus, suggested a further correction on the estimation of the probability of IBS. They argued that, although different allele frequencies at the loci used would imply different probabilities of IBS, the same coancestry estimation should be obtained from all of them for a particular dyad of individuals. Their WEDS estimator (Weighted Equal Drift Similarity) try to equalise the increase in coancestry from the base population for all loci by estimating the probability of IBS at each locus in the following way:

$$\hat{S}_l = \frac{\sum_{i=1}^{n_l} p_i^2 - S_{\min}}{1 - S_{\min}}$$

where $S_{\min}$ = min($\sum_{i=1}^{n_l} p_i^2$; l = 1,..., L). In this way we set $\hat{S}_l = 0$ for the locus having the lowest expected similarity (S).

Table 6b summarises some of the multilocus MME estimators that have been proposed.

Comparisons of the precision for different MME estimators can be found, for example, in Van de Casteele et al., (2001), Toro et al., (2002), Wang (2002), Csilléry et al., (2006) and Pino-Querido et al., (submitted).

Table 6b. Different multilocus estimators with their implicit assumptions and corresponding weights

Single locus estimator	Assumption	Multilocus estimator	Reference
Li and Horvitz (1953)		$$\hat{f} = \frac{\sum_{l=1}^{N} \hat{f}_l}{N}$$	Toro et al., (2002)
Li and Horvitz (1953)	Correlation between loci = 1	$$\hat{f} = \frac{\sum_{l=1}^{N} f_{Ml} - \sum_{l=1}^{N}\sum_{i=1}^{n} l}{N - \sum_{l=1}^{N}\sum_{i=1}^{n} p_{li}^2}$$	Ritland (1996)
Ritland (1996)	Correlation between loci = 0 (wl = nl − 1)	$$\hat{f} = \frac{\sum_{l=1}^{N}\sum_{i=1}^{n} \dfrac{f_{Mli} p_{li}^2}{p_{li}^2}}{\sum_{l=1}^{N}(n_l - 1)}$$	Ritland (1996)
Lynch and Ritland (1999) $$w_l = \frac{(1 + S_{ij})(p_i + p_j) - 4 p_i p}{8 p_i p_j}$$		$$\hat{f} = \frac{1}{\sum_{l=1}^{N} w_l} \sum_{l=1}^{N} w_l \hat{f}_l$$	Lynch and Ritland (1999)
Queller and Goodnight (1989)		Sum of the numerators of single locus estimators divided by the sum of denominators	Toro et al., (2002)
$$w_l^{-1} = \frac{\sum_{i=1}^{n_l} p_i^2 \left(1 - \sum_{i=1}^{n_l} p_i^2\right)}{(1 - \hat{s}_l)^2}$$		$$\hat{f} = \frac{1}{\sum_{l=1}^{L} w_l} \sum_{l=1}^{L} w_l \frac{S_l -}{1 -}$$	Oliehoek et al., (2006)

5.2.4. Maximum Likelihood Estimators (MLE)

Another way to deal with the problem of the estimation of coancestry is to calculate the probability of observing the genotypes of the dyad under study ($\varnothing_i$ in the Milligan 2003 notation; see also Herbinger et al., 1997; Kalinowski et al.,

2006) depending on a particular type of relationship (Δ). This is actually the likelihood of Δ, $L(\Delta) = \Pr(\mathcal{A}_i|\Delta) = \Sigma \Pr(\mathcal{A}_i|S_j)\Delta_j$, being S_j each of the nine identity modes explained in a previous section. Recalling that these modes can be summarised into the coefficients k_0, k_1 and k_2, and knowing that the coancestry coefficient in non inbred populations is $f = \frac{1}{4} k_1 + \frac{1}{2} k_2$, we can calculate the likelihood of a particular degree of coancestry from the expressions on Table 7, which depend on the frequencies of each allele in the population (p_i, p_j, p_k, p_m).

When information on more than one locus is available, the maximum likelihood estimator will be obtained by multiplying the likelihoods of the different loci. A revision of the factors affecting the precision of the MLE estimators as well as a comparison with the performance of the MME estimators can be found in Milligan (2003).

5.2.5. Problems of the Pairwise Estimators

Several are the factors that influence the efficiency of pairwise estimators. One of them is that these methods assume Hardy-Weinberg (HW) and gametic equilibrium at all loci analyzed. Although not formally proven, all authors agree that the departure from these assumptions may yield biased estimates.

Another major problem both of the MME and MLE methods is that they require the knowledge of the allelic frequencies in the base or reference population (Wang 2002; Toro et al., 2002; Thomas 2005).

Table 7. Probability of the nine possible combinations of genotypes in a dyad given identity by descent modes k_0, k_1, k_2.

X	Y	k_0	k_1	k_2
$A_i A_i$	$A_i A_i$	p_i^4	p_i^3	p_i^2
$A_i A_i$	$A_i A_j$	$2p_i^3 p_j$	$p_i^2 p_j$	
$A_i A_i$	$A_j A_j$	$p_i^2 p_j^2$		
$A_i A_j$	$A_i A_i$	$2p_i^3 p_j$	$p_i^2 p_j$	
$A_i A_j$	$A_i A_j$	$4 p_i^2 p_j^2$	$p_i p_j (p_i + p_j)$	$2 p_i p_j$
$A_i A_i$	$A_j A_k$	$2 p_i^2 p_j p_k$		
$A_j A_k$	$A_i A_i$	$2 p_i^2 p_j p_k$		
$A_i A_j$	$A_i A_k$	$4 p_i^2 p_j p_k$	$p_i p_j p_k$	
$A_i A_j$	$A_k A_m$	$4 p_i^2 p_j p_k p_m$		

When the number of generations elapsed from the base is small and the population size large, no relevant changes in the frequencies are expected and the estimates obtained from the individuals we actually want to determine their coancestries will be acceptable.

But if there had been enough margin for the genetic drift to act (many generations and/or reduced population sizes) frequencies could have changed and even some alleles lost in the process. Under this scenario, as shown by Toro et al., (2002), underestimations are systematically obtained, because of the magnified probability of IBS of the surviving alleles. MME methods may even provide negative values (out of the definition range, as it is a probability) making difficult the interpretation of the results. The correlation between estimates and true values of coancestry is low when wrong frequencies are used in the estimation, even with a large number of markers genotyped. When the true frequencies are known the performances of the estimators greatly improves (Toro et al., 2002; Pino-Querido et al., submitted).

In practical scenarios, the estimation of the coancestry between individuals reduces to choose between a few types of fixed relationships. For example, we may want to decide if two individuals are full-sibs or unrelated. The MLE methods, due to its nature, can deal with this situation just by looking for the degree of relationship with the highest likelihood and classifying the dyad in that group. If the appropriate choice is between two particular relationships (usually comparing a degree of relationship against being unrelated) a likelihood ratio can be calculated. The level of significance can be obtained by simulating a great number of dyads of the less likely relationship (using the population allelic frequencies) and then calculating the likelihood ratio for each of them under both hypotheses. The percentage of replicates whose value is above the observed ratio will be the probability of misclassification of the dyad (p-value).

Contrarily, the MME methods provide estimates on a continuous range. Therefore, it is necessary to determine the thresholds where assignation will change between fixed relationships (Figure 9). If no *a priori* information exists, values for those thresholds are usually the midpoints between actual genealogical values of each type of relation (Blouin et al., 1996), for example: 0.0625 between unrelated and half-sibs and 0.1875 between half-sibs and full-sibs (see Figure 9 and 10). The reason for these figures is to equalise the probabilities of under- and overestimating coancestry (type I and type II errors). The expected genealogical values may be substituted by the mean values of a large number of estimates from simulated dyads according to the observed population frequencies.

When the true relationships to be obtained appear in different frequencies, a common situation in Aquaculture where unrelated is the most common relationship, thresholds should be shifted to the less represented category to ensure the symmetry in the classification errors (Pino-Querido et al., submitted). But the relative proportion of each type of relationship usually is unknown *a priori*. These authors have proposed two methods to estimate the proportion of dyads belonging to each category using the properties of the normal curve. With their approach, the thresholds could be shifted to balance type I and II errors, and the resulting relatedness classification would improve.

When pairwise estimates are transformed to fixed genealogical relationships these methods can be affected by another problem. The consideration of each dyad independently of the rest of individuals may lead to incongruous assignations and aberrant coancestry matrices (Thomas and Hill 2000; Fernández and Toro 2006). For example, molecular information could be consistent with individuals A and B being full-sibs and similarly for individuals B and C. But this does not guarantee that individuals A and C will be correctly classified as full-sibs. Rodríguez-Ramilo et al., (in press) testing different pairwise estimators in a turbot population comprising 560 individuals with known relationships and genotyped for 11 microsatellites, have observed that the proportion of non reciprocal triplets of full-sibs is more than 70% of the triplets constructed from the molecular information.

Thomas and Hill (2002) pointed out that inferring coancestries in a pairwise mode may lead to reconstruct full-sib groups incompatible with Mendelian segregation rules. For example, three individuals with genotypes AA, BB and CC, respectively, cannot be full-sibs at the same time, but in pairwise comparisons this possibility will not be discarded. In the study by Rodríguez-Ramilo et al., (2007) the proportion of such incorrect groups was 60% of the total number of full-sibs families reconstructed with more than three individuals.

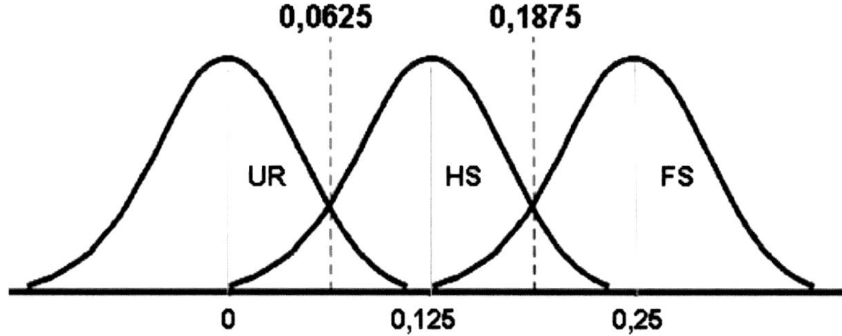

Figure 9. Distributions of the estimates for dyads of full-sibs (FS), half-sibs (HS) and unrelated (UR), respectively, with the thresholds commonly used for classification.

	R3	R5	R6	R8	R10	R13	R14	R19	R21	R23	R1	R4	R7
R3													
R5	-0.0260												
R6	0,1004	0,0907											
R8	0,0153	0,1548	0,1092										
R10	0,1146	0,0461	0,1256	0,1248									
R13	0,3352	0,1020	0,2131	-0.0015	0,1166								
R14	0,1028	0,2278	0,0287	0,2577	0,1031	0,0805							
R19	0,0898	0,1036	0,0603	0,3503	0,1186	0,0678	0,2052						
R21	-0.0305	0,0065	-0.0375	-0.0050	-0.0420	0,0444	0,1209	-0.0135					
R23	0,2921	0,0142	0,1009	-0.0915	0,0832	0,3354	-0.0555	0,0084	0,0033				
R1	0,2725	0,0419	0,0462	-0.1205	-0.0205	0,2891	-0.0090	-0.0250	0,0266	0,3012			
R4	0,0705	0,0335	-0.0475	0,0577	-0.0067	-0.0040	0,1415	0,0774	0,0548	-0.0150	-0.1035		
R7	-0.0265	-0.0700	-0.1070	0,0529	-0.0705	-0.0995	0,0056	0,0173	0,0487	-0.0880	-0.1210	0,0830	
R9	-0.0015	0,1077	0,0837	0,2460	0,0583	0,0098	0,1394	0,1845	0,1133	-0.0945	-0.0680	-0.0120	0,0093

Note.- Unrelated dyads are shown in green, half-sib dyads in yellow and full-sib dyads in red. The limits considered for classification were 0.0625 between unrelated and halfsibs, and 0.1875 between halfsibs and fullsibs.

Figure 10. Kinship classification of pairs of individuals (dyads) in a turbot broodstock

5.3. PEDIGREE RECONSTRUCTION

Methods grouped in this category were developed as an alternative to pairwise methods, to take advantage of the information from the rest of individuals on the population. A first approach was presented by Browning and Thompson (1999) who proposed to work with triplets of individuals instead of pairs. In this way, some genotype incompatibilities could be detected and coancestries assigned with more precision. A refinement of the method was developed by Sieberts et al., (2002) for joint estimation of triplets with mixed full-sibs, half-sibs, twin or unrelated, using Markov Chains to determine the posterior probability of each type of relationship.

Generalising, the ideal procedure would use the information from all the individuals in the population and would estimate every relationship simultaneously. Under this approach what we obtain is the global structure of relationships and, thus, the explicit reconstruction of the genealogy (at least for one generation). The negative side of these methods is that they require more complex calculations being computationally more demanding. Consequently, most of them have been limited to simple population structures like full-sib families.

Almudevar and Field (1999) presented an enumeration algorithm for feasible full-sib families, being the final objective to find the families with the highest number of members. When two alternate configurations comprising the same number of individuals exist, the one with the highest partial likelihood is chosen. Another method to obtain solutions involving the largest families is maximising the *Simpson's index* (Butler et al., 2004) defined for a population with N individuals structured in g groups comprising N_i elements as:

$$\frac{\sum_{i=1}^{g} N_i (N_i - 1)}{N(N-1)}$$

In this method, as in Almudevar and Field (1999), the compatibility of the proposed families with Mendelian segregation rules (e.g., all members could arise from a single couple) has to be tested to make sure they yield congruent solutions. The search for the optimum configuration is performed in both methods by Monte Carlo techniques.

In the literature, other estimators can be found relying on MCMC methods to search across the space of feasible solutions (compatible full-sib groups) but using different criteria to determine the more likely configuration. The driving parameter may be a combination of the likelihood ratios of all possible pairs of individuals (Smith et al., 2001) or the joint likelihood of each of the possible configurations (Painter 1997; Thomas and Hill 2000; Emery et al., 2001; Smith et al., 2001). The likelihood calculation of a particular family will depend, as in the pairwise MLE methods, on the allelic frequencies and the number of individuals of each genotype grouped together, as shown in Table 8.

Thus, a family with five of its members with genotype AA and another five with genotype BB, although possible, is much less likely than a family comprising heterozygous as well.

Thomas and Hill (2002) and Wang (2004b) extended these methods to populations with a hierarchical structure involving full-sib families and nested half-sib families.

Table 8. Likelihood of full-sibs families depending on the observed genotypes for one locus

Genotypes	Likelihood
AA	$p_A^4 + 4\,p_A^3\,p_X\,(1/2)^n + 4\,p_A^2\,p_X^2\,(1/4)^n$
AB	$2[p_A^2\,p_B^2 + (1/2)^{nAB}2(p_A^3\,p_B + p_A^2\,p_B\,p_X + p_A\,p_B^3 + p_A\,p_B^2\,p_X) + (1/4)^{nAB}\,4(p_A^2\,p_B\,p_X + p_A\,p_B^2\,p_X + p_A\,p_B\,p_X^2)] + (1/2)^{nAB}4\,p_A^2\,p_B^2$
AA BB	$4\,p_A^2\,p_B^2\,(1/4)^n$
AA AB	$4\,p_A^3\,p_B\,(1/2)^n + 4\,p_A^2\,p_B^2\,(1/4)^{nAA}(1/2)^{nAB} + 8\,p_A^2\,p_B\,p_X(1/2)^n$
AA AB BB	$4\,p_A^2\,p_B^2\,(1/4)^{nAA+nBB}(1/2)^{nAB}$
AA BC	$8\,p_A^2\,p_B\,p_C(1/4)^n$
AB AC	$4\,p_A^2\,p_B\,p_C(1/2)^n + 8\,p_A^2\,p_B\,p_C(1/4)^n + 8\,p_A\,p_B\,p_C\,p_D(1/4)^n$
AA AB AC	$8\,p_A^2\,p_B\,p_C(1/4)^n$
AA AB BC	$8\,p_A^2\,p_B\,p_C(1/4)^n$
AB AC BC	$8\,p_A^2\,p_B\,p_C(1 - p_X)(1/4)^n$
AA AB AC BC	$8\,p_A^2\,p_B\,p_C(1/4)$
AC BD	$8\,p_A\,p_B\,p_C\,p_D(1/4)^n$
AD AC BC	$8\,p_A\,p_B\,p_C\,p_D(1/4)^n$
AC AD BC BD	$8\,p_A\,p_B\,p_C\,p_D(1/4)^n$

p_X is the frequency of all alleles not present in the family considered

nAB is the number of individuals in a family carrying genotype *AB* (from a total of n full-sibs)

Although the pedigree reconstruction methods have clear advantages over the pairwise methods, they still present some drawbacks. First, its use is limited to very simple structures (full- or half-sibs) and requires more or less detailed knowledge of the population structure. Second, the likelihood calculations need the estimation of the true frequencies on the population (except for the estimator based on the Simpson's Index and the one by Almudevar and Field 1999, based on the partial likelihoods). Moreover, all formulas are also developed under the assumption of the existence of HW equilibrium for the alleles at each locus and linkage equilibrium between loci.

Recently, Fernández and Toro (2006) have developed a new method based on the observation by Toro et al., (2002) that, although pairwise molecular coancestry is an overestimation of the true value, the molecular coancestry matrix may exhibit a high correlation with the genealogical coancestry matrix. Consequently, these authors proposed a method which implements a *simulated annealing* algorithm to search within the space of all possible configurations of the population using the correlation between the molecular coancestry matrix and the genealogical matrix corresponding to the proposed structure, as the decision criterion. Obviously, only the compatible full-sibs families are considered. The advantage of this method is that it does not require the knowledge of the allelic frequencies in the reference population, and it makes no assumptions about the equilibrium within and between loci. Moreover, as calculations are simple, it allows for the estimation of more complex structures (e.g. non nested families) or even higher degree relationships like cousins or brothers from inbred parents, by reconstructing genealogies of more than one generation. The efficiency of some of the pedigree reconstruction estimators has been tested under different scenarios by Butler et al., (2004) and Fernández and Toro (2006).

A novel way of dealing with the problem of the joint estimation of all coancestry relationships was proposed by Beyer and May (2003). Their method is based on graphs theory representing the population as nodes (each individual) which can be linked by lines if they are full-sibs. Figure 11 shows different groups of connected nodes representing full-sibs families.

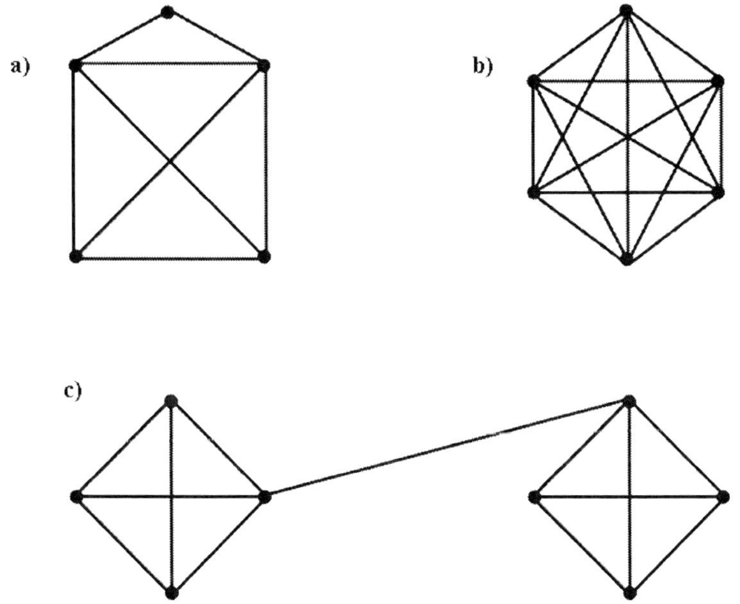

Modified from Beyer and May (2003)
a) Improper linked component (some individuals are not connected)
b) Proper/complete linked component (all elements connected with each other)
c) A wrong connection between two proper linked components (families).

Figure 11. Graphical representation of groups of individual and their sib-ship relation (linked points).

From an initial configuration based on the pairwise likelihood ratios (Goodnight and Queller 1999), we have to move to another structure where all individuals of the same family are connected to each other but disconnected from the nodes of other families, through graphical algorithms like the *minimum-cut*.

A final strategy to define groups of relatives consists in using the matrix of genetic distances (or relatedness coefficients) between individuals (e.g. allele sharing coefficient) to construct the subsequent phylogenetic trees using different algorithms (Bentzen et al., 2001). Those individuals clustering together will be assumed to belong to the same family. These methods, as suggested by Blouin (2003), appear to work reasonably well in situations where simple pedigree data are analyzed.

5.4. DOMINANT MARKERS

All previously explained estimators (except for the first similarity indices) have been developed for codominant markers. However, there are situations where it is much easy and cheaper to apply dominant markers, especially for species with low genomic resources. Therefore, it would be a good alternative to estimate coancestry from such type of markers.

Hardy (2003) proposed an estimator based on the fact that coancestry between two individuals is proportional to the correlation of the phenotypic frequencies for each allele in both individuals. For dominant loci those frequencies can only be 0 (lack of band) or 1 (presence of band, dominant allele present).

The advantage of this estimator, besides the nature of the markers it uses, is that it can be applied to populations not in HW equilibrium. However, to make use of this feature, the degree of deviation must be known (we have to calculate, through an independent estimate, the mean inbreeding coefficient of the population).

Table 9. Probability of the three possible combinations of phenotypes in a dyad given *identity by descent modes* k_0, k_1, k_2.

X	Y	k_0	k_1	k_2
Recessive	Recessive	q^4	q^3	q^2
Dominant	Recessive	$2(1-q^2)q^2$	$2q^2 - 2q^3$	
Dominant	Dominant	$2(1-q^2)^2$	$1 - 2q^2 + q^3$	$1 - q^2$

q is the frequency of the recessive allele.

Wang (2004a) and Ritland (2005) developed MME estimators from the expected values of the number of IBD alleles shared by two individuals (k_2, k_1, k_0) conditional on the phenotype of the dyad for the marker (Table 9). This table is equivalent to Table 7 with the difference that now the only three possibilities are both individuals presenting band, only one individual presenting band, or both individuals without band.

From these probabilities the reasoning is similar to that used for the MME methods on codominant markers (dependence on the knowledge of the frequencies, equilibrium within- and between-loci required, …).

5.5. GENOTYPING ERRORS

A major problem that cannot be denied is that inferences in any of the studied estimators are performed on molecular data that we assume correct. But molecular information may contain errors. The types of common errors were explained in a previous section. Some estimates of the rate of appearance of different kind of errors can be found in Talbot et al., (1995) and Ewen et al., (2000).

In the case of pairwise estimators, the incorrect assignation of a genotype to an individual affects their molecular similarity with the rest, but its influence will be relative depending on the amount of information from the other loci correctly genotyped. Epstein et al., (2000) included in their maximum likelihood estimator the possibility of errors in the genotype, assuming that when identical alleles are observed in two individuals the probability of that similarity corresponding to an error is ε and, consequently, the probability of both alleles being really equal would be $(1 - \varepsilon)$. Then, probabilities of being IBD conditional on the observed genotypes can be recalculated. Obviously, an estimate of the error rate has to be available.

However, methods involving pedigree reconstruction are more sensitive to point errors, as they may cause the individual bearing the erroneous genotype to be incompatible with its actual family. This effect is independent on the amount of error-free markers we have genotyped. In fact, the larger the number of loci genotyped, the higher the probability of including errors. A possible solution to the problem was proposed by Fernández and Toro (2006). The idea is relaxing the compatibility restrictions in the creation of full-sibs groups allowing for a number of loci where incompatibilities can exist. The expectation is that individuals with errors can be included in their correct family, at the risk of accepting erroneous assignations of individuals correctly genotyped.

Wang (2004b) extended a model similar to the one by Epstein et al., (2000) to calculate the likelihoods of different configurations of the population incorporating the probabilities of different type of errors (actually two types are distinguished). From the configuration with the highest likelihood, this author proposed to calculate the likelihood ratios for each family and locus assuming that one of the error rates is zero or both are null. In this way, it is possible to detect those loci more prone to bear an error and the more frequent kinds of error.

The use of any *a priori* information (e.g. genealogy) may be crucial to purge data before implementing any estimator. This practice will greatly improve the accuracy of the estimators.

5.6. MEASURES OF ACCURACY

The selection of the estimator to be used, besides the considerations regarding its nature (pairwise or not, dependent on frequencies, …), requires any measure of the accuracy or reliability which allows for comparison between different methods. These parameters can be used when genealogical data are also available or when simulation studies are conducted.

If we consider the coancestry as a value associated to each dyad, it seems appropriate to use the bias (difference between the estimated value and the true value) to determine the 'goodness' of the estimator. Although the expected value provided by an estimator may be unbiased, a great dispersion of estimates around the expectation may turn it into a low efficient estimator. One way to measure dispersion is the mean squared error (or its squared root; Milligan 2003).

Another proposed measure of fit is the correlation between the estimated coancestry matrix and the genealogical relationships for all possible dyads in the population (Toro et al., 2002; Fernández and Toro 2006). High correlation values mean that, although estimates may be biased, the ranking of pairs remains constant and also the relative values for different dyads.

When relationships between individuals are classified in a few types of coancestry (commonly full-sibs, half-sibs or non-related), accuracy of the method can be measured through the proportion of dyads correctly assigned to their actual group and the different errors of classification (Blouin et al., 1996; Thomas and Hill 2002; Fernández and Toro 2006). Figure 12 shows the nine possible combinations of estimated and real relationships when only three types are considered.

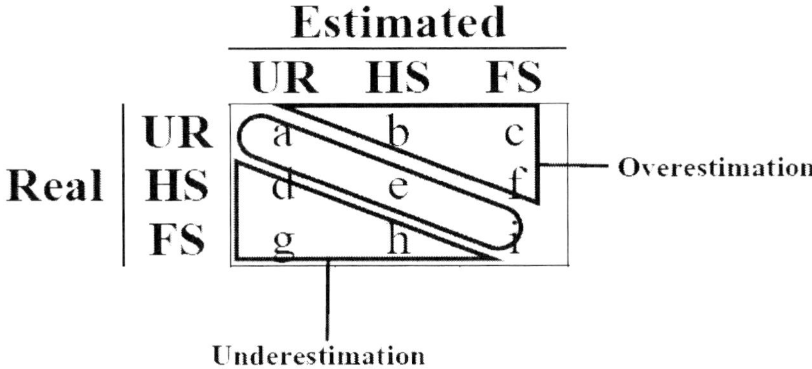

Figure 12. All types of possible assignations considering three kind of relationships: unrelated (UR), half-sibs (HS) or full-sibs (FS).

Using this kind of tables the tendency of a method to underestimate or overestimate coancestry can be determined, as well as its power for the detection of particular degrees of relationship. Obviously, before using this measure of fitness on MME pairwise estimators, we should transform the continuous values to qualitative data with the associated drawbacks already pointed out.

A more general comparison can be done through a Chi-square test, where the number of estimated and actual full-sibs, half-sibs and unrelated dyads are combined to calculate the statistic. The greater the value obtained, the worse the fit between estimates and true coancestries.

Smith et al., (2001), in the case of full-sibs family reconstruction, proposed a measure of accuracy which consisted in determining the lowest number of movements (i.e., exchanges of individuals between families) required to convert the estimated population structure into the true one. A refinement of this method distinguishes between movement of single individuals and the movement of blocks of individuals. For example: it is not equivalent to erroneously classify N individuals into N different families than splitting a correct family into two, one of the groups containing the N individuals.

Notwithstanding, as stated by different authors (Fernández and Toro 2006; Oliehoek et al., 2006), the appropriate comparison should be made in practical grounds, i.e.,: coancestry overestimates may be less important for the arrangement of the mating scheme under the minimum coancestry strategy, while underestimates are preferred if the coancestries have to be included in the estimation of breeding values using BLUP methodology. Oliehoek et al., (2006) found out that the ranking of estimators, according to the levels of genetic diversity preserved in a population managed under minimum coancestry contributions, changed depending on the population structure but not with the number of markers used or their polymorphism. In the aquaculture field, Porta and Fernández (unpublished data) have found that differences in accuracy between some coancestry estimators nearly vanished when values are used to organise individuals of a stock into different tanks as to yield the minimum average within-tank coancestry (mating between relatives is avoided and the genetic representation of each tank increases).

5.7. SOFTWARE

A large number of computer applications devoted to the estimation of coancestry from molecular information are available. Most of them implement pairwise methodologies, although several tools have been already developed for the use of group estimation. Table 10 summarises the most relevant programs and their characteristics.

Table 10. Computer applications devoted to the estimation of coancestry from molecular information

Program	Drawbacks / Comments	Reference	Web site
Pairwise (ML)			
RELPAIR	Allows for linkage / Accounts for genotyping errors / Allows for chromosome X markers/ Optional input of probable genealogies (checking pedigree errors)	Epstein et al., (2000)	http://www.sph.umich.edu/statgen/boehnke/relpair.html
PREST	Allows for linkage / Accounts for genotyping errors (some cases) / Optional input of probable genealogies (checking pedigree errors)	McPeek and Sun (2000)	http://galton.uchicago.edu/~mcpeek/software/prest/
ECLIPSE (in PANGAEA)	Analyses triplets of individuals / Allows for linkage / Accounts for genotyping errors (some cases) / Optional input of probable genealogies (checking pedigree errors)	Sieberts et al., (2002)	http://www.stat.washington.edu/thompson/Genepi/pangaea.shtml
ML-RELATE		Kalinowski et al., (2006)	http://www.montana.edu/kalinowski/kalinowski_software.htm
Pairwise (MM)			

Table 10. (Continued)

Program	Drawbacks / Comments	Reference	Web site
RELATEDNESS	*Only for Macintosh / Assumes unlinked loci*	*Queller and Goodnight (1989)*	*http://www.gsoftnet.us/GSoft.html*
KINSHIP	*Only for Macintosh / Assumes unlinked loci*	*Goodnight and Queller (1999)*	*http://www.gsoftnet.us/GSoft.html*
Program	*Drawbacks / Comments*	*Reference*	*Web site*
DELRIOUS	**Requires software Mathematica / Assumes unlinked loci**	**Stone and Björklund (2001)**	**http://www.zoo.utoronto.ca /stone/delrious/delrious.htm**
MER	**Dominant or codominant markers**	**Wang (2002); Wang (2004a)**	**http://www.zoo.cam.ac.uk/ioz/software.htm**
SPAGEDI	**Dominant or codominant markers / Compares genetic differentiation with geographical distance**	**Hardy and Vekemans (2002); Hardy (2003)**	**http://www.ulb.ac.be/sciences/ecoevol/spagedi.html**
Genealogy reconstruction			
BOREL (in PANGAEA)	**FS families (likelihood evaluation)**	**Painter (1997)**	**http://www.stat.washington .edu/thompson/Genepi/pangaea.shtml**
PRT	**FS families (enumeration of compatible groups)**	**Almudevar and Field (1999)**	**http://www.urmc.rochester .edu/smd/biostat/people/faculty/almudevar.html**
PARENTAGE	**FS and non-nested HS families / Allows for previous information / Accounts for genotyping errors / Slow and little accurate with large populations**	**Emery et al., (2001)**	**http://www.mas.ncl.ac.uk/ ~nijw/**
COLONY	**FS families nested in HS families / Accounts for genotyping errors / Accounts for known parents genotypes**	**Wang (2004b)**	**http://www.zoo.cam.ac.uk/ioz/software.htm**
MOLCOAN	**Not dependent on frequencies knowledge / FS and non-nested HS families (1 generation) or other relationships (< 1 generation) / Accounts for known genealogy / Accounts for genotyping errors**	**Fernández and Toro (2006)**	**http://www.uvigo.es/webs/c 03/webc03/XENETICA/X B2/Jesus/Fernandez.htm**

5.8. KINSHIP ESTIMATION IN AQUACULTURE

Molecular estimation of kinship in aquaculture has been mostly applied to evaluate family structure within a broodstock or at specific generations in a breeding programme to avoid inbreeding or loss of genetic diversity across generations of selection (Table 11). The main drawback for this approach is the large error associated to the estimators with the number of microsatellites commonly applied (around 10), which determines only enough power to discriminate with reasonable accuracy between quite divergent relationships (unrelated *vs.* full-sibs). This is obviously an important objective when beginning a genetic breeding programme with a broodstock of unknown origin, and probably, the application of a broader battery of microsatellites could be profitable to increase the accuracy of the estimations. The uncertainty associated to this analysis is on the basis of an important amount of works devoted to evaluate the accuracy of microsatellites comparing molecular estimations with actual pedigree data. As in paternity analysis kinship has been mainly estimated in fish and only using microsatellites.

Regarding the methodology, at all cases pairwise estimators were applied and usually categorical classification was made by applying thresholds between predetermined categories. The more recent development of pedigree reconstruction methods, and specially, their higher complexity for computation has probably limited their application. The Queller and Goodnight (1989) estimator of relatedness is the most popular and the programs by these authors, Relatedness and Kinship, have been widely applied. Pairwise distance matrices has also been used to reveal kinship through graphical methods using different algorithms for dendrogram reconstruction (UPGMA, Neigbor-joining, ..).

Table 11. Representative references for kinship evaluation in Aquaculture

Species	Taxonomic group	Pairwise estimator	Program	Analysis	Reference
Sparus aurata	Fish	r_{QG}	RELATED NESS	Family structure in broodstock	Castro et al., in press
Pangasianodo n gigas	Fish	r_R	Own program	Family structure in broodstock	Sriphairoj et al., 2007
Salvelinus alpinus	Fish	r_{QG}	KINSHIP	Family structure in broodstock	Ditlecadet et al., 2007
Solea senegalensis	Fish	r_{QG}	KINSHIP; GENETIX[4]	Family structure in broodstock and F1	Porta et al., 2006

Table 11. (Continued)

Species	Taxonomic group	Pairwise estimator	Program	Analysis	Reference
Solea senegalensis	Fish	r_{QG}	RELATED NESS	Family structure in broodstock	Castro et al., 2006
Haliotis discus hannai	Molluscs	r_{LR}	IDENTIX[2]	Assessing programmes for population enhancement	Sekino et al., 2005
Paralichthys olivaceus	Fish	r_{QG}	RELATED NESS; PHYLIP[2]	Heritability estimation through relatedness estimation	Shikano 2005
Scophthalmus maximus	Fish	r_{QG}	RELATED NESS	Accuracy of relatedness estimates *vs* real data	Borrell et al., 2004
Haliotis spp.	Molluscs	r_{QG}	RELATED NESS	Family structure in broodstock *vs* natural populations	Evans et al., 2004
Acipenser transmontanus	Fish	*r* for dominant markers[3]	Own program; PHYLIP[1]	Evaluation of microsatellites as dominant markers for estimating relatedness	Rodzen et al., 2004
Paralichthys olivaceus	Fish	r_{QG}	Own program	Using relatedness for enhancing genetic diversity in breeding programmes	Sekino et al., 2004
Oncorhynchus mykiss	Fish	r_{QG}	RELATED NESS	Accuracy of relatedness estimates *vs* real data	McDonald et al., 2004
Haliotis discus hannai	Molluscs	*D*	PHYLIP[1]	Family structure in broodstock and F1	Park and Kijima 2003
Salmo salar	Fish	r_{QG}	KINSHIP	Accuracy of relatedness estimates *vs* real data	Norris et al., 2000

The references are a representative sample taken relevant publications in Aquaculture

All studies were performed using microsatellite loci

[1]Clustering methods from relatedness matrices to identify full-sib families (PHYLIP program: Felsenstein, 1993).

[2]Belkhir et al., (2002)

[3]Lynch and Milligan (1994)

[4]Belkhir et al., (2003)

Pairwise estimators: QG (Queller and Goodnight, 1989); R (Ritland, 1996); LR (Lynch and Ritland , 1999); D (genetic distance)

ACKNOWLEDGEMENTS

Many examples and data presented in this article have been obtained by the Group of Genetics for Aquaculture and Conservation of the University of Santiago de Compostela (ACUIGEN). Our acknowledgement to Jaime Castro, Ania Pino-Querido, Miguel Hermida, Belén G. Pardo and Carmen Bouza, and to the technicians Lucía Insua, Sonia Gómez, María Portela, Susana Sánchez y María López.

REFERENCES

Almudevar A. and Field C. (1999) Estimation of single-generation sibling relationships based on DNA markers. *Journal of Agricultural Biological and Environmental Statistics* 4, 136–165.

Amos W. (1999). A comparative approach to the study of microsatellite. In: *Microsatellites. Evolution and Applications* (ed. by D.B. Goldstein and C. Schlötterer), pp. 66-79. Oxford University Press, Oxford.

Anderson E.C. and Garza J.C. (2006) The power of single-nucleotide polymorphisms for large-scale parentage inference. *Genetics* 172, 2567-2582.

Angers B. and Bernatchez L. (1997) Complex evolution of a salmonid microsatellite locus and its consequences in inferring allelic divergence from size information. *Molecular Biology and Evolution* 14, 230-238.

Araki H. and Blouin M.S. (2005) Unbiased estimation of relative reproductive success of different groups: evaluation and correction of bias caused by parentage assignment errors. *Molecular Ecology* 14, 4097-4109.

Ardren W.R., Borer S., Thrower F., Joyce J.E. and Kapuscinski A.R. (1999) Inheritance of 12 microsatellite loci in *Oncorhynchus mykiss. Journal of Heredity* 90, 529-536.

Bacon A.L., Dunlop M.G. and Farrington S.M. (2001) Hypermutability at a poly(A/T) tract in the human germline. *Nucleic Acids Research* 29, 4405-4413.

Ballou J.D. and Lacy R.C. (1995) Identifying genetically important individuals for management of genetic variation in pedigreed populations. In: *Population Management for Survival and Recovery*, (ed. by J.D. Ballou, M. Gilpin and T.J. Foose), pp. 76-111. Columbia University Press, New York.

Belkhir K., Castric V. and Bonhomme F. (2002) IDENTIX, a software to test for relatedness in a population using permutation methods. *Molecular Ecology Notes* 2, 611-614.

Belkhir K., Borsa P., Chikhi L., Raufaste N. and Bonhomme F. (2003) *GENETIX 4.04, logiciel sous Windows TM pour la génétique des populations.* Laboratoire Génome, Populations, Interactions, CNRS UMR 5000, Université de Montpellier II, Montpellier, France (in French).

Bernardo J.M. and Smith A.F.M. (1994) *Bayesian Theory.* John Wiley and Sons.

Bernatchez L. and Duchesne P. (2000) Individual-based genotype analysis in studies of parentage and population assignment: how many loci, how many alleles? *Canadian Journal of Fisheries and Aquatic Sciences* 57, 1-12.

Beyer J. and May B. (2003) A graph-theoretic approach to the partition of individuals into full-sib families. *Molecular Ecology* 12, 2243-2250.

Blouin M.S. (2003) DNA-based methods for pedigree reconstruction and kinship analysis in natural populations. *Trends in Ecology and Evolution* 18, 503-511.

Blouin M.S., Parsons M., Lacaille V. and Lotz S. (1996) Use of microsatellite loci to classify individuals by relatedness. *Molecular Ecology* 5, 393–401

Borrell Y.J., Alvarez J., Vázquez E., Pato C.F., Tapia C.M., Sánchez J.A. and Blanco G. (2004) Applying microsatellites to the management of farmed turbot stocks (*Scophthalmus maximus* L.) in hatcheries. *Aquaculture* 241, 133-150.

Boudry P., Collet B., Cornette F., Hervouet V. and Bonhomme F. (2002) High variance in reproductive success of the Pacific oyster (*Crassostrea gigas,* Thunberg) revealed by microsatellite-based parentage analysis of multifactorial crosses. *Aquaculture* 204, 283-296.

Bouza C., Castro J., Presa P., Sánchez L. and Martínez P. (2002) Allozyme and microsatellite diversity in natural and domestic populations of turbot (*Scophthalmus maximus*) in comparison with other Pleuronectiformes. *Canadian Journal of Fisheries and Aquatic Sciences* 59, 1460-1473.

Bowcock A.M., Ruiz-Linares A., Tomfohrde J., Minch E., Kidd J.R. and Cavalli-Sforza L.L. (1994) High resolution of human evolutionary trees with polymorphic microsatellites. *Nature* 368, 455-457.

Brookfield J.F.Y. (1996) A simple new method for estimating null allele frequency from heterozygote deficiency. *Molecular Ecology* 5, 453-455.

Brown R.C., Tsalavouta M., Terzoglou V., Magoulas A. and Mcandrew B.J. (2005) Additional microsatellites for *Sparus aurata* and cross-species amplification within the Sparidae family. *Molecular Ecology Notes* 5, 605-607.

Browning S. and Thompson E.A. (1999) Interference in the analysis of genetic marker data. *American Journal of Human Genetics* Suppl 65, A244

Butler K., Field C., Herbinger C.M. and Smith B.R. (2004) Accuracy, efficiency and robustness of four algorithms allowing full sibship reconstruction from DNA marker data. *Molecular Ecology* 13, 1589–1600.

Caballero A. and Toro M.A. (2000) Interrelations between effective population size and other pedigree tools for the management of conserved populations. *Genetical Research* 75, 331-343.

Carreras-Carbonell J., Macpherson E. and Pascual M. (2004) Isolation and characterization of microsatellite loci in *Tripterygion delaisi*. *Molecular Ecology Notes* 4, 438-439.

Castro J., Bouza C., Sánchez L., Cal R.M., Piferrer F. and Martínez P. (2003) Gynogenesis assessment by using microsatellite genetic markers in turbot (*Scophthalmus maximus*). *Marine Biotechnology* 5, 584-592.

Castro J., Bouza C., Presa P., Pino-Querido A., Riaza A., Ferreiro I., Sánchez L. and Martínez P. (2004) Potential sources of error in parentage assessment of turbot (*Scophthalmus maximus*) using microsatellite loci. *Aquaculture* 242, 119-135.

Castro J., Pino-Querido A., Chavarrías D., Merino P., Ferreiro I., Riaza A., Hermida M., Bouza C., Sánchez L. and Martínez P. (2005) Asignación de parentescos mediante marcadores microsatélite en rodaballo (*Scophthalmus maximus*), lenguado (*Solea* senegalensis) y dorada (*Sparus aurata*). pp. 456-457 *X Congreso Nacional de Acuicultura*. Gandía, Valencia.

Castro J., Pino A., Hermida M., Bouza C., Riaza A., Ferreiro I., Sánchez L. and Martínez P. (2006) A microsatellite marker tool for parentage analysis in Senegal sole (*Solea senegalensis*): Genotyping errors, null alleles and conformance to theoretical assumptions. *Aquaculture* 261, 1194-1203.

Castro J., Pino A., Hermida M., Bouza C., Chavarrías D., Merino P., Sánchez L. and Martínez P. (2007) A microsatellite marker tool for parentage assessment in gilthead sea bream (*Sparus aurata*). *Aquaculture* 272S1, S210-S216.

Cercueil A., Bellemain E. and Manel S. (2002) PARENTE: computer program for parentage analysis. *Journal of Heredity* 93, 458-459.

Chakraborty R., Meagher T.R. and Smouse P.E. (1988) Parentage analysis with genetic markers in natural populations. I. The expected proportion of offspring with unambiguous paternity. *Genetics* 118, 527-536.

Chistiakov D.A., Hellemans B. and Volckaert F.A.M. (2006) Microsatellites and their genomic distribution, evolution, function and applications: A review with special reference to fish genetics. *Aquaculture* 255, 1-29.

Csilléry K., Johnson T., Beraldi D., Clutton-Brock T., Coltman D., Hansson B., Spong G. and Pemberton J.M. (2006) Performance of marker-based relatedness estimators in natural populations of outbred vertebrates. *Genetics* 173, 2091–2101

Dakin E.E. and Avise J.C. (2004) Microsatellite null alleles in parentage analysis. *Heredity* 93, 504-509.

Danzmann R.G. (1997) PROBMAX: a computer program for assigning unknown parentage in pedigree analysis from known genotypic pools of parents and progeny. *Journal of Heredity* 88, 333.

De León F.J.G., Canonne M., Quillet E., Bonhomme F. and Chatain B. (1998) The application of microsatellite markers to breeding programmes in the sea bass, *Dicentrarchus labrax*. *Aquaculture* 159, 303-316.

Ditlecadet D., Dufresne F., Le Francois N.R. and Blier P.U. (2006) Applying microsatellites in two commercial strains of Arctic charr (*Salvelinus alpinus*): Potential for a selective breeding program. *Aquaculture* 257, 37-43.

Dong S.R., Kong J., Zhang T.S., Meng X.H. and Wang R.C. (2006) Parentage determination of Chinese shrimp (*Fenneropenaeus chinensis*) based on microsatellite DNA markers. *Aquaculture* 258, 283-288.

Duchesne P., Godbout M.H. and Bernatchez L. (2002) PAPA (package for the analysis of parental allocation): a computer program for simulated and real parental allocation. *Molecular Ecology Notes* 2, 191-193.

Duchesne P., Castric T. and Bernatchez L. (2005) PASOS (parental allocation of singles in open systems): a computer program for individual parental allocation with missing parents. *Molecular Ecology Notes* 5, 701-704.

Ellegren H. (2000) Microsatellite mutations in the germline: implications for evolutionary inference. *Trends in Genetics* 16, 551-558.

Emery A.M., Wilson I.J., Craig S., Boyle P.R. and Noble L.R. (2001) Assignment of paternity groups without access to parental genotypes: multiple mating and developmental plasticity in squid. *Molecular Ecology* 10, 1265–1278

Emik L.O. and Terrill C.E. (1949) Systematic procedures for calculating inbreeding coefficients. *Journal of Heredity* 40, 51–55.

Epstein M.P., Duren W.L. and Boehnke M. (2000) Improved inference of relationship for pairs of individuals. *American Journal of Human Genetics* 67, 1219–1231

Estoup A., Gharbi K., San Cristobal M., Chevalet C., Haffray P. and Guyomard R. (1998) Parentage assignment using microsatellites in turbot (*Scophthalmus maximus*) and rainbow trout (*Oncorhynchus mykiss*) hatchery populations. *Canadian Journal of Fisheries and Aquatic Sciences* 55, 715-725.

Evans B., Bartlett J., Sweijd N., Cook P. and Elliott N.G. (2004) Loss of genetic variation at microsatellite loci in hatchery produced abalone in Australia (*Haliotis rubra*) and South Africa (*Haliotis midae*). *Aquaculture* 233, 109-127.

Ewen K.R., Bahlo M., Treloar S.A., Levinson D.F., Mowry B., Barlow J.W. and Foote S.J. (2000) Identification and analysis of error types in highthroughput genotyping. *American Journal of Human Genetics* 67, 727–736

Ezaz M.T., Sayeed S., McAndrew B.J. and Penman D.J. (2004) Use of microsatellite loci and AFLP markers to verify gynogenesis and clonal lines in Nile tilapia *Oreochromis niloticus* L. *Aquaculture Research* 35, 1472-1481.

Falconer D.S. and Mackay T.F. (1996) *Introduction to Quantitative Genetics*. 4th ed. Longman, Harlow, Essex.

Felsenstein J. (1993) *PHYLIP (Phylogeny Inference Package) version 3.5c.* Distributed by the author. Department of Genetics, University of Washington, Seattle.

Fernández J. and Toro M.A. (2006) A new method to estimate relatedness from molecular markers. *Molecular Ecology* 15, 1657–1667.

Fessehaye Y., El-Bialy Z., Rezk M.A., Crooijmans R., Bovenhuis H. and Komen H. (2006) Mating systems and male reproductive success in Nile tilapia (*Oreochromis niloticus*) in breeding hapas: a microsatellite analysis. *Aquaculture* 256, 148-158.

Fiumera A.C., DeWoody Y.D., DeWoody J.A., Asmussen M.A. and Avise J.C. (2001) Accuracy and precision of methods to estimate the number of parents contributing to a half-sib progeny array. *Journal of Heredity* 92, 120–126.

Fopp-Bayat D., Kolman R. and Woznicki P. (2007) Induction of meiotic gynogenesis in sterlet (*Acipenser ruthenus*) using UV-irradiated bester sperm. *Aquaculture* 264, 54-58.

Gelfand A.E. and Smith A.F.M. (1990) Sampling based approaches to calculating marginal densities. *Journal of the American Statistical Association* 85, 398-409.

Gerber S., Mariette S., Streiff R., Bodenes C. and Kremer A. (2000) Comparison of microsatellites and amplified fragment length polymorphism markers for parentage analysis. *Molecular Ecology* 9, 1037-1048.

Gjedrem T. (2005) *Selection and Breeding Programs in Aquaculture*. Springer, Dordrecht.

Goodnight K. F. and Queller D.C. (1999) Computer software for performing likelihood tests of pedigree relationship using genetic markers. *Molecular Ecology* 8, 1231–1234.

Hadfield J.D., Richardson D.S. and Burke T. (2006) Towards unbiased parentage assignment: combining genetic, behavioural and spatial data in a Bayesian framework. *Molecular Ecology* 15, 3715-3730.

Hara M. and Sekino M. (2003) Efficient detection of parentage in a cultured Japanese flounder *Paralichthys olivaceus* using microsatellite DNA marker. *Aquaculture* 217, 107-14.

Hardy O. (2003) Estimation of pairwise relatedness between individuals and characterization of isolation-by-distance processes using dominant genetic markers. *Molecular Ecology* 12, 1577–1588.

Hardy O. and Vekemans X. (2002) SPAGeDi: a versatile computer program to analyse spatial genetic structure at the individual or population levels. *Molecular Ecology Notes* 2, 618–620.

Hastein T., Hill B.J., Berthe F. and Lightner D.V. (2001) Traceability of aquatic animals. *Rev. Sci. Tech. – Off. Int. Épizoot.* 20, 564-583.

Hayes B., Sonesson A.K. and Gjerde B. (2005) Evaluation of three strategies using DNA markers for traceability in aquaculture species. *Aquaculture* 250, 70-81.

Herbinger C.M., Doyle R.W., Pitman E.R., Paquet D., Mesa K.A., Morris D.B., Wright J.M. and Cook D. (1995) DNA fingerprint based analysis of paternal and maternal effects on offspring growth and survival in communally reared rainbow trout. *Aquaculture* 137, 245-256.

Herbinger C.M., Doyle R.W., Taggart C.T., Lochman S.E., Brooker A.L., Wright J.M. and Cook D. (1997) Family relationships and effective population size in a natural cohort of Atlantic cod (*Gadus morhua*) larvae. *Canadian Journal of Fisheries and Aquatic Sciences* 54, 11–18.

Hoffman J.I. and Amos W. (2005) Does kin selection influence fostering behaviour in Antarctic fur seals (*Arctocephalus gazella*)? *Proceedings of the Royal Society B-Biological Sciences* 272: 2017-2022.

Hulata G. (2001) Genetic manipulations in aquaculture: a review of stock improvement by classical and modern technologies. *Genetica* 111, 155-173.

Iyengar A., Piyapattanakorn S., Heipel D.A., Stone D.M., Howell B.R., Child A.R. and MaClean N. (2000) A suite of highly polymorphic microsatellite markers in turbot (*Scophthalmus maximus* L.) with potential for use across several flatfish species. *Molecular Ecology* 9, 368-371.

Jerry D.R., Preston N.P., Crocos P.J., Keys S., Meadows J.R.S. and Li Y.T. (2004) Parentage determination of Kuruma shrimp *Penaeus (Marsupenaeus) japonicus* using microsatellite markers (Bate). *Aquaculture* 235, 237-247.

Jerry D.R., Evans B.S., Kenway M. and Wilson K. (2006a) Development of a microsatellite DNA parentage marker suite for black tiger shrimp *Penaeus monodon. Aquaculture* 255, 542-547.

Jerry D.R., Preston N.P., Crocos P.J., Keys S., Meadows J.R.S. and Li Y.T. (2006b) Application of DNA parentage analysis for determining relative growth rates of *Penaeus japonicus* families reared in commercial ponds. *Aquaculture* 254, 171-181.

Jin L. and Chakraborty R. (1995) Population structure, stepwise mutations, heterozygote deficiency and their implications in DNA forensics. *Heredity* 74, 274–285.

Johnson N.A., Rexroad C.E., Hallerman E.M., Vallejo R.L. and Palti Y. (2007) Development and evaluation of a new microsatellite multiplex system for parental allocation and management of rainbow trout (*Oncorhynchus mykiss*) broodstocks. *Aquaculture* 266, 53-62.

Jones A.G. (2001) GERUD 1.0: a computer program for the reconstruction of parental genotypes from progeny arrays using multilocus DNA data. *Molecular Ecology Notes* 1, 215-218.

Jones A.G. (2005) GERUD 2.0: a computer program for the reconstruction of parental genotypes from half-sib progeny arrays with known or unknown parents. *Molecular Ecology Notes* 5, 708-711.

Jones A.G. and Ardren W.R. (2003) Methods of parentage analysis in natural populations. *Molecular Ecology* 12, 2511-2523.

Jones A.G., Kvarnemo C., Moore G.I., Simmons L.W. and Avise J.C. (1998) Microsatellite evidence for monogamy and sex-biased recombination in the Western Australian seahorse, *Hippocampus angustus. Molecular Ecology* 7, 1497-1505.

Kalinowski S.T., Wagner A.P. and Taper M.L. (2006) ML-RELATE: a computer program for maximum likelihood estimation of relatedness and relationship. *Molecular Ecology Notes* 6, 576–579.

Kalinowski S.T., Taper M.L. and Marshall T.C. (2007) Revising how the computer program CERVUS accommodates genotyping error increases success in paternity assignment. *Molecular Ecology* 16, 1099-1106.

Kasumovic M.M., Ratcliffe L.M. and Boag P.T. (2003) A method to improve confidence in paternity assignment in an open mating system. *Canadian Journal of Zoology* 81, 2073-2076.

Lemos A., Freitas A.I., Fernandes A.T., Goncalves R., Jesus J., Andrade C. and Brehm A. (2006) Microsatellite variability in natural populations of the blackspot seabream *Pagellus bogaraveo* (Brunnick, 1768): a database to access parentage assignment in aquaculture. *Aquaculture Research* 37, 1028-1033.

Li Q., Park C. and Kijima A. (2003) Allelic transmission of microsatellites and application to kinship analysis in newly hatched Pacific abalone larvae. *Fisheries Science* 69, 883-889.

Li Y.T., Wongprasert K., Shekhar M., Ryan J., Dierens L., Meadows j., Preston N., Coman G. and Lyons R.E. (2007) Development of two microsatellite multiplex systems for black tiger shrimp *Penaeus monodon* and its application in genetic diversity study for two populations. *Aquaculture* 266, 279-288.

Li C.C. and D.G. Horvitz (1953) Some methods of estimating the inbreeding coefficient. *American Journal of Human Genetics* 5, 107-117.

Li C.C., Weeks D.E. and Chakrabarti A. (1993) Similarity of DNA fingerprints due to chance and relatedness. *Human Heredity* 43, 45-52.

Liu Z.J. and Cordes J.F. (2004) DNA marker technologies and their applications in aquaculture genetics. *Aquaculture* 238, 1-37.

Lucas T., Macbeth M., Degnan S.M., Knibb W. and Degnan B.M. (2006) Heritability estimates for growth in the tropical abalone *Haliotis asinina* using microsatellites to assign parentage. *Aquaculture* 259, 146-152.

Luikart G. and England P.R. (1999) Statistical analysis of microsatellite DNA data. *Trends in Ecology and Evolution* 14, 253– 256.

Lynch M. and Ritland K. (1999) Estimation of pairwise relatedness with molecular markers. *Genetics* 152, 1753-1766.

Malécot G. (1948) *Les Mathématiques de l'Hérédité*. Masson et Cie, Paris.

Marshall T.C., Slate J., Kruuk L.E.B. and Pemberton J.M. (1998) Statistical confidence for likelihood-based paternity inference in natural populations. *Molecular Ecology* 7, 639-655.

McCartney M.A., Brayer K. and Levitan D.R. (2004) Polymorphic microsatellite loci from the red urchin, *Strongylocentrotus franciscanus*, with comments on heterozygote deficit. *Molecular Ecology Notes* 4, 226-228.

McDonald G.J., Danzmann R.G. and Ferguson M.M. (2004) Relatedness determination in the absence of pedigree information in three cultured strains of rainbow trout (*Oncorhynchus mykiss*). *Aquaculture* 233, 65–78.

McPeek M.S. and Sun L. (2000) Statistical tests for detection of misspecified relationships by use of genome screen data. *American Journal of Human Genetics* 66, 1076–1094.

Meagher T.R. (1986) Analysis of paternity within a natural population of *Chamaelirium luteum*. I. Identification of most-likely male parents. *American Naturalist* 128, 199-215.

Meagher T.R. and Thompson E.A. (1986) The relationship between single and parent pair genetic likelihoods in genealogy reconstruction. *Theoretical Population Biology* 29, 87–106.

Milkman R. and Zeitler R.R. (1974) Concurrent multiple paternity in natural and laboratory populations of *Drosophila melanogaster. Genetics* 78, 1191-1193.

Milligan B.G. (2003) Maximum-likelihood estimation of relatedness. *Genetics* 163, 1153–1167.

Neef B.D., Repka J. and Gross M.R. (2001) A Bayesian framework for parentage analysis: The value of genetic and other biological data. *Theoretical Population Biology* 59, 315-331.

Nielsen R., Mattila D.K., Clapham P.J. and Palsboll P.J. (2001) Statistical approaches to paternity analysis in natural populations and applications to the North Atlantic humpback whale. *Genetics* 157, 1673-1682.

Norris A.T., Bradley D.G. and Cunningham E.P. (2000) Parentage and relatedness determination in farmed Atlantic salmon (*Salmon salar*) using microsatellite markers. *Aquaculture* 182, 73-83.

Oliehoek P.A., Windig J.J., van Arendonk J.A.M. and Bijma P. (2006) Estimating relatedness between individuals in general populations with a focus on their use in conservation programs. *Genetics* 173, 483–496.

Painter I. (1997) Sibship reconstruction without parental information. *Journal of Agricultural Biological and Environmental Statistics* 2, 212-229.

Pardo B.G., Fernández C., Hermida M., Vázquez A., Pérez M., Presa P., Calaza M., Alvarez-Dios J.A., Comesaña A.S., Raposo-Guillán J., Bouza C. and Martínez P. (2007) Development and characterization of 248 novel microsatellite markers in turbot (*Scophthalmus maximus*). *Genome* 50, 329-332.

Pena S.D.J. and Chakraborty R. (1994) Paternity testing in the DNA era. *Trends in Genetics* 10, 204-209.

Pérez-Enríquez R., Takagi M. and Taniguchi N. (1999) Genetic variability and pedigree tracing of a hatchery-reared stock of red sea bream (*Pagrus major*) used for stock enhancement, based on microsatellite DNA markers. *Aquaculture* 173 411-421.

Pinera J.A., Bernardo D., Blanco G., Vázquez E. and Sánchez J.A. (2006) Isolation and characterization of polymorphic microsatellite markers in *Pagellus bogaraveo*, and cross-species amplification in *Sparus aurata* and *Dicentrarchus labrax. Molecular Ecology Notes* 6, 33-35.

Pompanon F., Bonin A., Bellemain E. and Taberlet P. (2005) Genotyping errors: causes, consequences and solutions. *Nature Reviews Genetics* 6, 847-859.

Porta J., Porta J.M., Martínez-Rodríguez G. and Alvarez M.C. (2006) Genetic structure and genetic relatedness of a hatchery stock of Senegal sole (*Solea senegalensis*) inferred by microsatellites. *Aquaculture* 251, 46-55.

Primmer C.R., Saino N., Moller A.P. and Ellegren H. (1998) Unravelling the processes of microsatellite evolution through analysis of germ line mutations in barn swallows *Hirundo rustica*. *Molecular Biology Evolution* 15, 1047-1054.

Queller D.C. and Goodnight K.F. (1989) Estimating relatedness using molecular markers. *Evolution* 43, 258-274.

Rengmark A.H., Slettan A., Skaala O., Lie O. and Lingaas F. (2006) Genetic variability in wild and farmed Atlantic salmon (*Salmon salar*) strains estimated by SNP and microsatellites. *Aquaculture* 253, 229-237.

Ritland K. (1996) Estimators for pairwise relatedness and inbreeding coefficients. *Genetical Research* 67, 175-186.

Ritland K. (2005) Multilocus estimation of pairwise relatedness with dominant markers. *Molecular Ecology* 14, 3157–3165.

Rodríguez-Ramilo S., Toro M.A., Martínez P., Castro J., Bouza C. and Fernández J. (2007) Accuracy of the pairwise methods in the reconstruction of family relationships using molecular information of turbot (*Scophthalmus maximus*). *Aquaculture*.273, 434-442.

Rodzen J.A., Famula T.R. and May B. (2004) Estimation of parentage and relatedness in the polyploid white sturgeon (*Acipenser transmontanus*) using a dominant marker approach for duplicated microsatellite loci. *Aquaculture* 232, 165-182.

Sancristobal M. and Chevalet C. (1997) Error tolerant parent identification from a finite set of individuals. *Genetical Research* 70, 53-62.

Schlötterer C. (2000) Evolutionary dynamics of microsatellite DNA. *Chromosoma* 109, 365-371.

Sekino M., Saido T., Fujita T., Kobayashi R. and Takami H. (2005) Microsatellite DNA markers of Ezo abalone (*Haliotis discus hannai*): a preliminary assessment of natural populations sampled from heavily stocked areas. *Aquaculture* 243, 33-47.

Sekino M., Saitoh K., Yamada T., Kumagai A., Hara M. and Yamashita Y. (2003) Microsatellite-based pedigree tracing in a Japanese flounder *Paralichthys olivaceus* hatchery strain: implications for hatchery management related to stock enhancement program. *Aquaculture* 221, 255-263.

Sekino M., Sugaya T., Hara M. and Taniguchi N. (2004) Relatedness inferred from microsatellite genotypes as a tool for broodstock management of Japanese flounder *Paralichthys olivaceus*. *Aquaculture* 233, 163-172.

Shikano T. (2005) Marker-based estimation of heritability for body color variation in Japanese flounder *Paralichthys olivaceus*. *Aquaculture* 249, 95-105.

Sieberts S.K., Wijsman E.M. and Thompson E.A. (2002) Relationship inference from trios of individuals, in the presence of typing error. *American Journal of Human Genetics* 70, 170-180

Signorovitch J. and Nielsen R. (2002) PATRI- paternity inference using genetic data. *Bioinformatics* 18, 341-342.

Smith B.R., Herbinger C.M. and Merry H.R. (2001) Accurate partition of individuals into full-sib families from genetic data without parental information. *Genetics* 158, 1329-1338

Sneath P.H.A. and Sokal R.R. (1973) *The Principles and Practice of Numerical Classifications*. W.H. Freeman, San Francisco.

Sriphairoj K., Kamonrat W. and Na-Nakorn U. (2007) Genetic aspect in broodstock management of the critically endangered Mekong giant catfish, *Pangasianodon gigas* in Thailand. *Aquaculture* 264, 36-46.

Stone J. and Björklund M. (2001) DELRIOUS: a computer program designed to analyse molecular marker data and calculate delta and relatedness estimates with confidence. *Molecular Ecology Notes* 1, 209–212.

Taggart J.B. (2007) FAP: an exclusion-based parental assignment program with enhanced predictive functions. *Molecular Ecology Notes* 7, 412-415.

Talbot C.C. Jr, Avramopouls D., Gerken S., Chakravarti A., Armour J.A., Matsunami N., White R. and Antonarakis S.E. (1995) The tetranucleotide repeat polymorphism D21S1245 demonstrates hypermutability in germline and somatic cells. *Human Molecular Genetics* 4, 1193-1199.

Taris N., Baron S., Sharbel T.F., Sauvage C. and Boudry P. (2005) A combined microsatellite multiplexing and boiling DNA extraction method for high-throughput parentage analyses in the Pacific oyster (*Crassostrea gigas*). *Aquaculture Research* 36, 516-518.

Thomas S.C. (2005) The estimation of genetic relationships using molecular markers and their efficiency in estimating heritability in natural populations. *Philosophical Transactions of the Royal Society of London. Series B, Biological Sciences* 360, 1457-1467.

Thomas S.C. and Hill W.G. (2000) Estimating quantitative genetic parameters using sibships reconstructed from marker data. *Genetics* 155, 1961–1972.

Thomas S.C. and Hill W.G. (2002) Sibship reconstruction in hierarchical population structures using Markov chain Monte Carlo techniques. *Genetical Research* 79, 227–234.

Thompson E.A. (1991) Estimation of relationships from genetic data. In: *Handbook of Statistics* (ed. by C.R. Rao and R. Chakraborty), vol. 8, pp. 255-269. Elsevier Science Publishers, Amsterdam.

Toro M.A., Barragán C., Óvilo C., Rodrigañez J., Rodríguez C. and Silió L. (2002) Estimation of coancestry in Iberian pigs using molecular markers. *Conservation Genetics* 3, 309-320.

van de Caastele T., Galbusera P. and Matthysen E. (2001) A comparison of microsatellite-based pairwise relatedness estimators. *Molecular Ecology* 10, 1539-1549.

Van Oosterhout C., Hutchinson W.F., Wills D.P.M. and Shipley P. (2004) micro-checker: software for identifying and correcting genotyping errors in microsatellite data. *Molecular Ecology Notes* 4, 535-538.

Vandeputte M., Kocour M., Mauger S., Dupont-Nivet M., De Guerry D., Gela D., Vallod D., Linhart O. and Chevassus B. (2005) Heritability estimates for growth-related traits using microsatellite parentage assignment in juvenile common carp (*Cyprinus carpio* L.). *Aquaculture* 247, 31-32.

Vandeputte M., Mauger S. and Dupont-Nivet M. (2006) An evaluation of allowing for mismatches as a way to manage genotyping errors in parentage assignment by exclusion. *Molecular Ecology Notes* 6, 265-267.

Volckaert F.A.M. and Hellemans B. (1999) Survival, growth and selection in a communally reared multifactorial cross of African catfish (*Clarias gariepinus*). *Aquaculture* 171, 49-64.

Wang X.X., Ross K.E., Saillant E., Gatlin D.M. and Gold J.R. (2007) Genetic effects on carcass-quality traits in hybrid striped bass (*Morone chrysops* female x *Morone saxatilis* male). *Aquaculture Research* 38, 973-980.

Wang J. (2002) An estimator for pairwise relatedness using molecular markers. *Genetics* 160, 1203–1215.

Wang J. (2004a) Estimating pairwise relatedness from dominant genetic markers. *Molecular Ecology* 13: 3169–3178.

Wang J. (2004b) Sibship reconstruction from genetic data with typing errors. *Genetics* 166, 1963-1979.

Wang J. (2006) Informativeness of genetic markers for pairwise relationship and relatedness inference. *Theoretical Population Biology* 70, 300-321.

Wang X.X., Ross K.E., Saillant E., Gatlin D.M. and Gold J.R. (2006) Quantitative genetics and heritability of growth-related traits in hybrid striped bass (*Morone chrysops* female x *Morone saxatilis* male). *Aquaculture* 261, 535-545.

Weir B.S., Anderson A.D. and Heppler G.M. 2006 Genetic relatedness analysis: modern data and new challenges. *Nature Reviews Genetics* 7, 771-780.

Wesmajervi M.S., Westgaard J.I. and Delghandi M. (2006) Evaluation of a novel pentaplex microsatellite marker system for paternity studies in Atlantic cod (*Gadus morhua* L.). *Aquaculture Research* 37, 1195-1201.

Wilson A.J. and Ferguson M.M. (2002) Molecular pedigree analysis in natural populations of fishes: approaches, applications, and practical considerations. *Canadian Journal of Fisheries and Aquatic Sciences* 59, 1696-1707.

Worthington Wilmer J., Allen P.J., Pomeroy P.P., Twiss S.D. and Amos W. (1999) Where have all the fathers gone? An extensive microsatellite analysis of paternity in the grey seal (*Halichoerus grypus*). *Molecular Ecology* 8, 1417-1429.

Zane L., Bargelloni L. and Patarnello T. (2002) Strategies for microsatellite isolation: a review. *Molecular Ecology* 11, 1-16.

INDEX

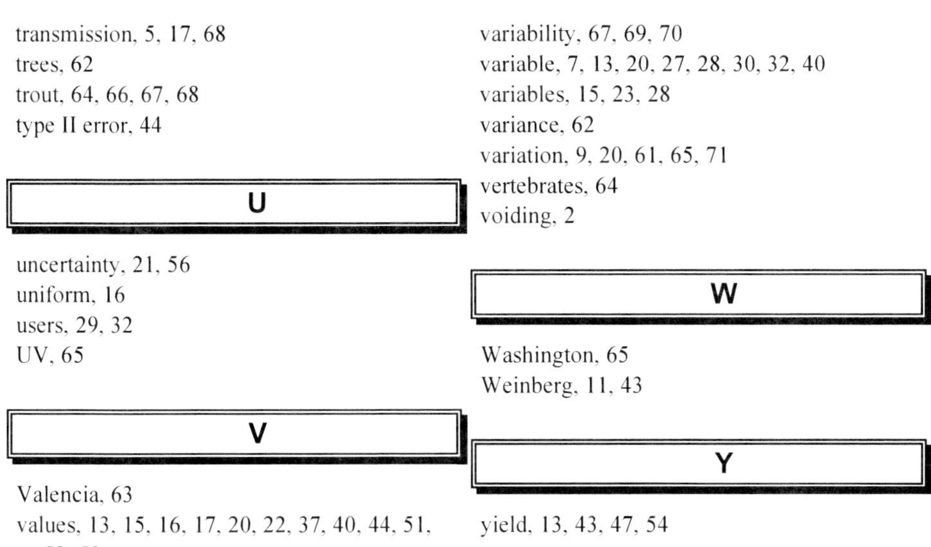